Progress in Optical Science and Photonics

Volume 14

The purpose of the series Progress in Optical Science and Photonics is to provide a forum to disseminate the latest research findings in various areas of Optics and its applications. The intended audience are physicists, electrical and electronic engineers, applied mathematicians, biomedical engineers, and advanced graduate students.

More information about this series at http://www.springer.com/series/10091

Aavishkar Katti · R. A. Yadav

Optical Spatial Solitons in Photorefractive Materials

Springer

Aavishkar Katti
Dr. Vishwanath Karad MIT World Peace University
Pune, Maharashtra, India

R. A. Yadav
Banaras Hindu University
Varanasi, Uttar Pradesh, India

ISSN 2363-5096 ISSN 2363-510X (electronic)
Progress in Optical Science and Photonics
ISBN 978-981-16-2549-7 ISBN 978-981-16-2550-3 (eBook)
https://doi.org/10.1007/978-981-16-2550-3

This Springer imprint is published by the registered company Springer Nature Singapore Pte Ltd.
The registered company address is: 152 Beach Road, #21-01/04 Gateway East, Singapore 189721, Singapore

Aavishkar Katti would like to dedicate this book to,

Mom and Dad, who have been role models for me in my academic career,

The Omniscient Lord Almighty, Lord Dattatreya, who is within and without

Foreword

It gives me great pleasure to write the foreword for the book by Prof. Aavishkar Katti and Prof. R. A. Yadav.

Optical solitons, in general, have been researched since long in the past. In particular, photorefractive solitons or spatial solitons in photorefractive materials have been quite attractive in recent times given their potential for practical applications in optical networks, optical computing, navigation and waveguides.

This book is a timely addition to the literature of soliton optics. It acts as a bridge between the discovery of photorefractive solitons back in the 90s and the current research being carried out in this field. The book provides a theoretical approach to understand the soliton formation phenomena in photorefractive materials. The theoretical foundation of diverse types of solitons supported by photorefractive media has been laid out in terms of nonlinear partial differential equation systems. Wherever the system cannot be solved exactly, numerical methods are resorted to and a generalized mathematical formulation is built for the propagation and stability characteristics of photorefractive solitons in diverse configurations. The mathematical treatment for spatial photorefractive soliton starts from the simplest photorefractive material supporting screening solitons and then spans out to various different configurations including pyroelectric crystals, photorefractive crystal circuits and photorefractive waveguides.

The book will be useful for beginning researchers in the field of photorefractive solitons since it contains a logical and easily understandable chronological development of the theoretical formulation for the various characteristics of the solitons. Graduate students in nonlinear optics in particular and nonlinear dynamics in general can also benefit from this book.

I wish all success to the book, and hope that it will benefit the targeted readership.

R. K. Sharma.

Prof. (Dr.) R. K. Sharma
ConsenSys Block Chain Professor and Former Head
Department of Mathematics
Indian Institute of Technology Delhi
New Delhi, India

Preface

The present book has been an effort which has been fructified as a result of a sincere wish for advancement of research in nonlinear optical materials including photorefractive media. It has been a satisfying academic journey where there have been moments when we have fallen in love with the subject again.

When the photorefractive effect was discovered, it was generally thought to be damaging and unwanted because of the coupling away of energy by beam fanning. But soon, it was found that photorefractive effect has tremendous applications in holography, optical phase conjugation, wave mixing and optical storage. These processes were all driven by the diffusion effect induced in the photorefractive crystal due to the photogeneration of carriers. In the 1990s, it was discovered that photorefractive materials can support self-trapping, the exact opposite of the processes mentioned above. An external field contributing a drift component of the current of the photogenerated carriers could result in an index waveguide to form due to an incident beam of light. Photorefractive crystals are unique because they exhibit a saturable nonlinearity, and spatial solitons can be realized in the laboratory at relatively low laser powers. Certain characteristics of spatial solitons in general were discovered by studying the properties of spatial solitons in photorefractive crystals.

This treatise has been aimed at beginning researchers in the field of optical solitons. In particular, we have presented a detailed study of optical spatial solitons in a special class of nonlinear optical crystals, which are photorefractive crystals. In Chap. 1, a succinct but clear introduction of the photorefractive solitons has been presented in this chapter. The electro-optic effect has been explained with a relevant example which then leads to the explanation of the photorefractive effect and self-trapping in photorefractive media. A brief experimental overview has also been given after which we discuss a comprehensive theory for the simplest type of photorefractive solitons, known as screening solitons. The derivation of the induced space charge field is discussed, and the normalized intensity profiles are obtained along with the soliton width existence curves for all three types of solitons, viz. bright, dark and grey. The effect of the diffusion on the propagation of the soliton is investigated. The diffusion effect results in a self-deflection of the soliton trajectory which follows a parabolic curve now. Once the investigation of the simplest type of photorefractive soliton is completed, we move on to discuss optical spatial solitons in photorefractive crystals having a finite bulk photovoltaic coefficient and pyroelectric coefficient.

In Chap. 2, a theoretical formulation for optical spatial solitons in photovoltaic photorefractive crystals is studied. The bulk photovoltaic effect is different from the conventional "solar cell" photovoltaic effect and relates to anisotropic probabilities of electron processes leading to a giant photovoltage. The band transport model is used to obtain the space charge field and the dynamical evolution equation for light beams propagating in biased photovoltaic photorefractive crystals. The general formulation presented here branches out to systems of photorefractive crystal which is open-circuited or closed-circuited and without the external bias field. Relevant example is taken to show the dark and grey soliton states. Consequently, the theoretical formulation for self-trapping due to the pyroelectric effect is elucidated. The pyroelectric effect relates to a temperature change induced transient electric field. The temperature change can be externally controlled or due to incident beam energy absorption. This transient pyroelectric field can self-trap a soliton in a photorefractive crystal. The space charge field and the dynamical evolution equation are obtained using simple considerations of charge transport and continuity equations. Finally, the interplay between the pyroelectric field, photovoltaic field and the external bias field is discussed.

The stability of optical spatial solitons is a crucial aspect to be studied since only the stable optical spatial solitons have potential practical applications. In Chap. 3, we shall be studying the stability of photorefractive solitons by using the modulation instability theory. Modulation instability, also known as sideband instability, can be said to be a phenomenon where certain perturbations of a periodic waveform get strengthened consequently leading to the generation of spectral sidebands and the ensuing breakup of the waveform into small filaments. A generalized theory for the modulation instability gain for broad optical beams in photorefractive photovoltaic crystals is elucidated which can, in turn, be reduced to the cases of broad optical beams in biased and unbiased photovoltaic crystals, non-photovoltaic photorefractive crystals and centrosymmetric photorefractive crystals. Lastly, the theoretical formulation for the modulation instability gain for light beams in pyroelectric photorefractive crystals is expounded.

The next logical question that should be addressed is the formation of the photorefractive solitons. A theory for time dependence and formation of optical spatial solitons in photorefractive crystals have been discussed in Chap. 4. The temporal evolution of the induced space charge field is studied, and hence, the dynamical evolution equation with an explicit temporal dependence is derived. A general theory for incorporating the temporal characteristics in the theoretical formulation for photorefractive solitons is then discussed. This theory is illustrated to be valid for different types and configurations of photorefractive crystals. The ratio of the maximum intensity to the dark irradiance of the incident beam profoundly affects the temporal characteristics, and hence, the temporal development of the soliton width has been studied as a function of the intensity ratio. The relevant conditions for formation of steady state and quasi-steady-state solitons are studied. The magnitude of the electro-optic coefficients affects the temporal evolution of solitons which has been discussed. We then discuss how this theory can be reduced to study the temporal evolution of screening solitons and centrosymmetric solitons. In addition, the theory for temporal evolution of solitons in biased photovoltaic photorefractive crystals is also considered. A further reduction of this theory for screening solitons and photovoltaic solitons is briefly commented upon.

In Chap. 5, a comprehensive theory for coupled soliton pairs in photorefractive materials is presented for incoherent and coherent soliton pairs in various different configurations and types of photorefractive crystals. The two beams need to have the same polarization, wavelength and should be mutually incoherent for the coupled soliton pairs to be realized. There exist diverse realizations of the soliton pairs like bright-bright, dark-dark, grey-grey and bright-dark. A theory for coupled spatial solitons in photovoltaic photorefractive crystals is illustrated first. The space charge field is obtained and the conditions under which the system reduces to study open-circuit photovoltaic and closed-circuit photovoltaic spatial soliton pairs is elucidated. Subsequently, a theory for coupled spatial soliton pairs in non-photovoltaic photorefractive crystals is illustrated. This theory, in turn, is shown to reduce to the system studying screening soliton pairs and centrosymmetric soliton pairs under appropriate conditions. Coupled soliton pairs can be either incoherently coupled or coherently coupled depending upon the mutual phase relationship between the two light beams, and both of these are studied in detail. Lastly, a novel soliton pair known as Gaussian or quasi-soliton pair is investigated in photorefractive materials. The condition for existence of a vast set of such Gaussian or quasi-soliton pairs is delineated along with the pertinent parameter space. The properties and dynamical evolution characteristics of such Gaussian solitons pairs are studied in detail.

Another type of coupling observed in photorefractive crystal circuits is the separate coupling. In Chap. 6, we first delineate a general theory for optical spatial solitons propagating in a biased photorefractive crystal circuit. We shall then investigate the validity of the theory for various different types and configurations of the constituent photorefractive crystals. The approach to derive the coupled space charge fields in both the crystals of the photorefractive crystal circuit will be discussed. The intensity profile and propagation of spatial soliton in each crystal can affect the intensity profile and propagation of the spatial soliton in the other crystal. The input intensity of each soliton and the temperature of each crystal exert a coupling effect between the individual solitons in the two crystals. These phenomena are discussed in detail taking relevant examples. This type of separate coupling for the two solitons exists because of the light-induced current in each photorefractive crystal flowing through the circuit. A brief idea about further reading for the stabilities and dynamical evolution of such separately coupled solitons will be presented.

Finally, in Chap. 7, a general theory for the existence of bright solitons is studied in a photorefractive waveguide under the paraxial and Wentzel–Kramers–Brillouin–Jeffreys approximation. The planar waveguide structure increases the self-focussing while decreasing the minimum or threshold power required for self-trapping. The waveguide structure lowers the power required to form a soliton. The power required to just self-trap a soliton keeps on reducing as the strength of the waveguide increases.

Distinct regions of power are identified in which the characteristics of the self-trapping are discussed in detail. The propagation of the solitons is visualized with and without the presence of the waveguide structure. This theory can be applied to waveguides embedded in photorefractive crystals exhibiting the linear and quadratic electro-optic effect simultaneously, which then reduces to centrosymmetric photorefractive waveguides and conventional photorefractive waveguides. The interesting case of pyroelectric photorefractive waveguides is also discussed while photovoltaic waveguides are briefly commented upon.

In summary, we have tried to present a clear theoretical formulation for different types, configurations and characteristics of photorefractive solitons. We cannot claim that this book covers all the topics related to photorefractive solitons, since it is a vast field and further research in photorefractive solitons is going on at a fast pace. What this book does is provide a useful knowledge of the theory behind important aspects and characteristics of photorefractive solitons. The readers will find themselves equipped very robustly for pursuing further research in this area after a thorough reading.

Banasthali, Tonk, India

Aavishkar Katti
R. A. Yadav

Acknowledgements

The writing of this monograph has been a journey in which many people have helped in some way or the other. Even a word of encouragement in difficult circumstances can go a long way in ensuring continuity in academic success. The authors would firstly, like to acknowledge our Publishing Editor, Dr. Loyola D'Silva and the Project Co-ordinator, Mr. Prasanna Kumar Narayanasamy for their continuous support throughout the manuscript writing and submission process.

The authors would like to acknowledge fruitful discussions with Prof. C. P. Katti, Professor, School of Computers and System Sciences, Jawaharlal Nehru University, New Delhi. He has been instrumental in shaping the understanding of the mathematical foundation of optical solitons in nonlinear media. Various discussions with him on diverse concepts of Mathematical Physics have immensely helped in the writing of this treatise.

Discussions with Dr. Awadhesh Prasad, Associate Professor, Department of Physics and Astrophysics, University of Delhi, has also contributed immensely to our knowledge in various aspects of nonlinear dynamics.

Dr. Kaushik Chaudhury, Non-Executive Director, Iota design and innovations lab Pvt. Ltd. has been a support throughout the writing of this book. His never say die attitude and advice from time to time has been invaluable. Dr. Parvez Ahmed Alvi, Associate Professor, Department of Physics, Banasthali Vidyapith should find a special mention here for providing continuous encouragement. Last but not the least, Ms. Apurva Sharma must be acknowledged for her support without which writing this book would not have been possible.

Aavishkar Katti
R. A. Yadav

Contents

Chapter 1
Introduction to Photorefractive Solitons

In linear optics, the index of refraction n is a constant depending upon the frequency of light. The refraction and reflection are thus independent of the intensity of the light beam. The index of refraction of a medium is, in general dependent upon the distribution of electrons and arrangement of atoms. When a light beam illuminates a crystal, the charges are driven up and down by the electric field of the incident light. These moving electrons radiate light and generate an electric field proportional to the incident field. Hence, there appears a phase shift in the transmitted field which is equivalent to a slower propagation velocity. If the intensity of the field is small such that the electric field of the light beam is much lesser than the intra-atomic field, this phase shift will be independent of intensity. Therefore, the index of refraction is not dependent on the intensity of light. This is the regime of linear optics.

When the intensity of the light beam is comparable to the intra-atomic electric field, the electron distribution in the medium is modified by the incident radiation. Hence, the index of refraction is indeed dependent on the intensity of the incident beam. This is the regime of non-linear optics.

1.1 Electro-Optic Effect

The phenomenon of the electro-optic effect can be illustrated as a refractive index change consequent to the application of an external electric field. This effect is observed in solid or liquid medium with anisotropic properties. Electro-optic effect can be broadly said to be of two types, the Pockels' effect where the index change is proportional to the electric field and the Kerr effect, where the index change is proportional to the square of electric field. The Pockels effect is also known as the linear electro-optic effect while the Kerr effect is also known as the quadratic electro-optic effect. In noncentrosymmetric media, Pockel's effect is much stronger than Kerr effect. In centro-symmetric media (media with inversion symmetry), the Kerr Effect dominates. Mathematically, the electro-optic effect can be defined by expressing the

A. Katti and R. A. Yadav, *Optical Spatial Solitons in Photorefractive Materials*, Progress in Optical Science and Photonics 14, https://doi.org/10.1007/978-981-16-2550-3_1

change in the impermeability tensor [1].

$$\Delta\eta_{ij} = \Delta\left(\frac{1}{n^2}\right)_{ij} = r_{ijk}E_k + s_{ijkm}E_kE_m \tag{1.1}$$

where E_k, E_m are the components of the applied electric field. The first term on the RHS of (1.1) represents the Pockels' effect while the second term represents the Kerr effect. The change in the dielectric tensor $\Delta\epsilon$ is related to the change in the impermeability tensor $\Delta\eta$ by,

$$\Delta\epsilon = -\frac{\epsilon\Delta\eta\epsilon}{\epsilon_0} \tag{1.2}$$

Moving to a principal co-ordinate system, in which the dielectric tensor is diagonal, the change can be expressed as,

$$\Delta\epsilon_{ij} = -\epsilon_0 n_i^2 n_j^2 \Delta\eta_{ij} \tag{1.3}$$

Two examples, namely GaAs and $LiNbO_3$ will be considered to illustrate the mathematical formulation of the electro-optic effect.

1.2 Illustration of the Electro-Optic Effect: GaAs and $LiNbO_3$

GaAs has only the linear electro-optic effect. So $s_{ijkm} = 0$. The linear electro-optic coefficients r_{ijk} are components of a tensor of rank 3. So there are 9 components of this tensor. However, the symmetry properties of ϵ and η reduce the independent components to 18. Also, the symmetry leads to an interchangeability in i and j,

$$r_{ijk} = r_{jik}$$

Due to this above mentioned symmetry, contracted indices will used henceforth defined by,

$$1 = (11) = (\text{xx}), 2 = (22) = (\text{yy}), 3 = (33) = (\text{zz}), 4 = (23) = (32) = (\text{yz}) = (\text{zy})$$
$$5 = (31) = (13) = (\text{zx}) = (\text{xz}), 6 = (12) = (21) = (\text{xy}) = (\text{yx})$$

Using the contracted notation, we can express the electro-optic coefficients in terms of a 6×3 matrix. It is important to realize here that these matrices do not have the usual transformation property of a regular tensor. Although there are 18 independent components of the linear electro-optic coefficients, many of these are zero or identical due to point group symmetry in many of the photorefractive crystals.

GaAs has a point group. The electro-optic coefficients in contracted notation are given by [2],

$$\begin{bmatrix} 0 & 0 & 0 \\ 0 & 0 & 0 \\ 0 & 0 & 0 \\ r_{41} & 0 & 0 \\ 0 & r_{41} & 0 \\ 0 & 0 & r_{41} \end{bmatrix} \tag{1.4}$$

where $r_{51} = r_{63} = r_{41}$. Let the applied electric field be $\vec{E} = \left(E_x, E_y, E_z\right)$. The change in the impermeability tensor can be written as,

$$\Delta\eta = \begin{bmatrix} 0 & r_{41}E_z & r_{41}E_y \\ r_{41}E_z & 0 & r_{41}E_x \\ r_{41}E_y & r_{41}E_x & 0 \end{bmatrix} \tag{1.5}$$

and the corresponding change in the dielectric tensor will be,

$$\Delta\epsilon = -\epsilon_0 n^4 \begin{bmatrix} 0 & r_{41}E_z & r_{41}E_y \\ r_{41}E_z & 0 & r_{41}E_x \\ r_{41}E_y & r_{41}E_x & 0 \end{bmatrix} \tag{1.6}$$

where n denotes the index of refraction of the medium.

The index ellipsoid in contracted notation becomes,

$$\left(\frac{1}{n_x^2} + r_{1k}E_k\right)x^2 + \left(\frac{1}{n_y^2} + r_{2k}E_k\right)y^2 + \left(\frac{1}{n_z^2} + r_{3k}E_k\right)z^2 + 2yzr_{4k}E_k$$
$$+ 2zxr_{5k}E_k + 2xyr_{6k}E_k = 1 \tag{1.7a}$$

($k = 1, 2, 3$).

Using the equation (1.7a) for the index ellipsoid in contracted notation and (1.4), the index ellipsoid can be written,

$$\frac{x^2}{n^2} + \frac{y^2}{n^2} + \frac{z^2}{n^2} + 2yzr_{41}E_x + 2zxr_{41}E_y + 2xyr_{41}E_z = 1 \tag{1.7b}$$

The cross terms are the terms containing xz, xy and yz. It can be inferred that the ellipsoid's major axes do not coincide with the x, y, and z crystal axes. Choosing the case when the electric field is along the z-axis, the equation of the index ellipsoid now is,

$$\frac{x^2}{n^2} + \frac{y^2}{n^2} + \frac{z^2}{n^2} + 2xyr_{41}E_z = 1 \tag{1.8}$$

The equation (1.8) can be diagonalized by

$$\begin{aligned} x &= x'\cos 45° - y'\sin 45° \\ y &= x'\sin 45° - y'\cos 45° \\ z' &= z \end{aligned} \tag{1.9}$$

Substituting (1.9) into (1.8) gives,

$$\frac{x'^2}{n'^2_x} + \frac{y'^2}{n'^2_y} + \frac{z'^2}{n'^2_z} \tag{1.10}$$

with

$$n'_z = n \tag{1.11}$$

and

$$\begin{aligned} \tfrac{1}{n'^2_x} &= \tfrac{1}{n^2} + r_{41}E_z \\ \tfrac{1}{n'^2_y} &= \tfrac{1}{n^2} - r_{41}E_z \end{aligned} \tag{1.12}$$

Assuming that $n^2 r_{41} E_z << 1$, the new principal indices of refraction are,

$$\begin{aligned} n_{x'} &= n - \tfrac{1}{2}n^3 r_{41}E_z \\ n_{y'} &= n + \tfrac{1}{2}n^3 r_{41}E_z \end{aligned} \tag{1.13}$$

where use has been made of the following equation,

$$\Delta n = -\frac{1}{2}n^3 \Delta\left(\frac{1}{n^2}\right) \tag{1.14}$$

In the case of an $LiNbO_3$ crystal, the electro-optic coefficients in contracted notation are [1],

$$\begin{bmatrix} 0 & r_{12} & r_{13} \\ 0 & -r_{12} & r_{13} \\ 0 & 0 & r_{33} \\ 0 & r_{42} & 0 \\ r_{42} & 0 & 0 \\ r_{12} & 0 & 0 \end{bmatrix} \tag{1.15}$$

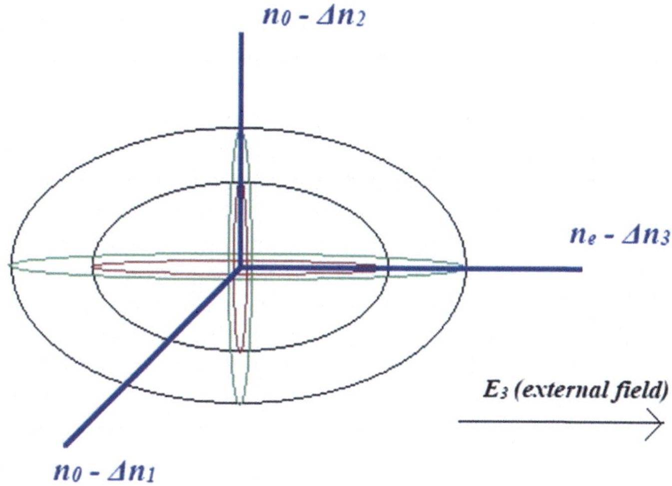

Fig. 1.1 The modified index ellipsoid for the case of $LiNbO_3$ crystal

Let the applied electric field be $\vec{E} = (E_x, E_y, E_z)$. Choosing the case when the electric field is along the z-axis, the equation of the index ellipsoid now is,

$$\left(\frac{1}{n_0^2} + r_{13}E_z\right)x^2 + \left(\frac{1}{n_0^2} + r_{13}E_z\right)y^2 + \left(\frac{1}{n_e^2} + r_{33}E_z\right)y^2 \tag{1.16}$$

Hence,

$$\Delta\left(\frac{1}{n_z^2}\right) = -2\frac{1}{n_e^3}\Delta(n_1) = r_{13}E_z \tag{1.17}$$

$$\Delta(n_z) = -\frac{n_e^3}{2}r_{33}E_z \tag{1.18}$$

The index ellipsoid is modified as shown in Fig. 1.1.

1.3 Photorefractive Effect

The photorefractive effect, which is a combination of the electro-optic effect and photoconductivity has many applications in the fields of optoelectronics and photonics [3–6]. This effect was first discovered in 1966 as a degradation of a linearly polarized laser beam propagating in a plate of $LiNbO_3$ by Ashkin et al. [7]. The interesting observation here was that no change was observed for beams polarized perpendicular to the c-axis, i.e., ordinary waves while the extraordinary waves or the beams polarized parallel to the c-axis suffered a spreading along the same axis. This resulted in a broadened light spot on the observation screen. It can be reasonably inferred that this phenomenon involved a light-induced refractive index change due to the incident beam.

The photorefractive effect is an optical nonlinearity dependent on the induction of a space charge field. The space charge field is formed due to transport of photogenerated charge carriers by diffusion or drift mechanisms and the consequent recombination. Photorefraction can be said to be a combination of the electro-optic effect and photoconductivity. The photorefractive nonlinearity cannot be represented in terms of *m*-order susceptibility $\chi(m)$. A theory which introduces an effective susceptibility has been proposed [8], but it has very little practical use in photorefractive optics as we are more concerned about the functional form of the refractive index change induced by the intensity of the incident light beam. The magnitude of the photorefractive effect is dependent on the magnitude of the electro-optic coefficients because these electro-optic coefficients are responsible for converting the induced space charge field into a quantifiable refractive index change. Since the photogeneration and recombination occurs at the donors and acceptors, so the concentration of the defects and impurities is another factor which influences the magnitude of the photorefractive effect. Under suitable conditions, the photorefractive effect can be observed for intensities as low as mW/cm^2. There is one drawback of the photorefractive nonlinearity that is its slow response time. The response time is related to the time for buildup of the space charge field. Hence the photorefractive response time is related to the carrier mobility and is typically of the order of the dielectric response time. The induced photorefractive damage can be present for varying amounts of time depending upon the photorefractive material. The induced space charge field persists in the dark for about microseconds in case of semiconductors while this can extend upto months and years in case of single crystals like $LiNbO_3$. The photorefractive effect is reversible and the space charge field can be erased by an incident beam causing intense uniform illumination. To summarize, the photorefractive nonlinearity is nonlocal, depends upon the doping and imperfections in the materials, has a high sensitivity, low response time, and the ensuing space charge field is erasable by homogeneous illumination.

1.3.1 Physical Mechanism: Standard Microscopic Model

The mechanism behind the photorefractive effect can be explained lucidly by considering the Fig. 1.2. The electro-optic material has both donor and acceptor centers denoted by N_D and N_A respectively. These may be the two valence states of the same impurity or defect. When a light beam with a nonuniform spatial intensity distribution illuminates a photorefractive crystal, the donor centres are ionized and photogenerated carriers arise. These photogenerated carriers can be either electrons or holes. Due to the nonuniform concentration gradient of charge carriers across the spatial dimension of the PR crystal, a diffusion current is set up. In addition, there can be a drift component also depending upon an externally applied electric field and

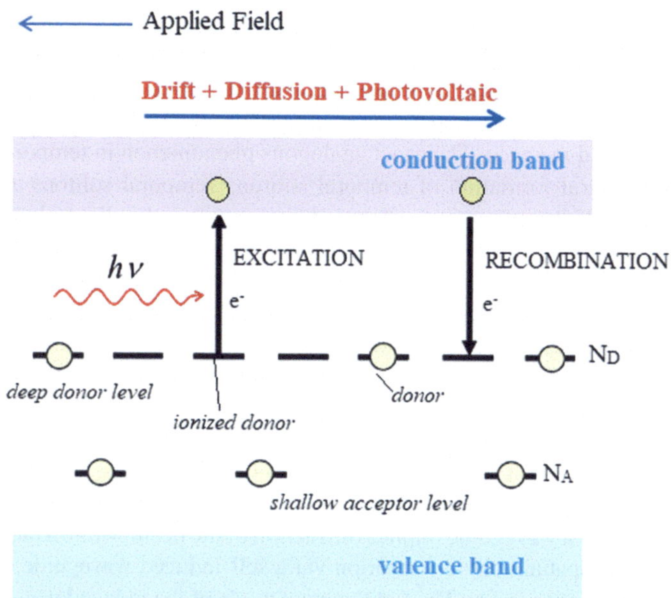

Fig. 1.2 Graphical representation of the process of charge redistribution leading to photorefraction

a finite photovoltaic or pyroelectric coefficient. The photovoltaic and pyroelectric component of drift current will be discussed later in this chapter. Now, these free charge carriers move through the conduction band and then recombine at the acceptors. There is an ensuing charge redistribution between the regions of illumination and the dark regions which creates a space charge field. This space charge field causes a refractive index change through the electro-optic effect. We shall study the mathematical formulation detailing the photorefractive charge transport in subsequent sections.

1.3.2 Photorefractive Solitons

Solitons, which can be defined to be solitary waves travelling unchanged for long distances find mention in diverse fields of physics, including nonlinear optics, plasma physics, string theory, hydrodynamics, statistical mechanics and biology [9–16]. The historical discovery of solitons took place in water and is very interesting. In 1834, near the Union Canal at Hermiston at the Herriott-Watt University, Edinburgh, John Scott Russell observed that a pile of water in a canal proliferated without any spreading or attenuation over several kilometers. His report, published in 1844, detailed his observations [17]. It is notable that such waves were originally known as *solitary waves*. The inverse scattering method is a powerful method to understand the theory behind these solitary waves in water [18, 19]. These solitary waves remain intact after mutual collisions and so, the term *soliton* has been coined for these waves because of their particle-like nature [10]. In nonlinear optics, solitons can be categorized as being either spatial or temporal. Spatial optical solitons result from an exact balancing of the diffraction by the nonlinearity. Hence, spatial solitons can be said to be confined in space. The exact analogous phenomenon in temporal domain can be said to heral formation of temporal soliton. Temporal solitons are optical pulses of light which propagate undistorted with time as the dispersion has been balanced exactly by the nonlinearity. The phenomenon of the electro-optic effect causes a nonlinear refractive change in the crystal being illuminated by an incident light beam [20–22]. Self focussing or self defocussing can occur due to the intensity dependence of the refractive index and temporal self-phase modulation. A spatial soliton is realized when the diffraction of the beam is balanced exactly by the self-focusing effects due to nonlinearity. The optical beam induces a dielectric waveguide which then guides itself. The refractive index variation in this induced waveguide would be much greater at the center of the beam than at its ends. The power of the incident beam defines whether or not self trapping will occur. Low enough power not capable of forming a waveguide implies diffraction of the beam, while a large enough intensity implies spatial soliton formation via a self-induced waveguide due to the change in the refractive index. The fundamental mode of the induced waveguide can be understood to be the spatial soliton itself. Such a nonlinear waveguide can even guide a weak probe beam of a different frequency or polarization [23]. In analogy, it is the self phase modulation which is crucial in formation of temporal solitons since it opposes the dispersion induced broadening of the light pulse [24].

A lens analogy can also be taken to understand the formation of spatial solitons. Diffraction causes the wavefront of the incoming beam to become curved and hence the beam spreads to a wide region. This is like a concave lens curving the wavefront. Now, the self focusing induces a refractive index gradient which tries to focus the wavefront like a convex lens. When these two aforementioned lensing effects cancel out each other exactly, a stable self trapped beam can propagate and is known as a spatial soliton [24]. The spatial intensity profile should be of specific shapes so as to result in perfect and unchanging solitary wave. The set of such specific profiles can be said to be the nonlinear equivalent of the linear waveguide modes of self-induced refractive index variation. Optical spatial solitons can be of three types, namely bright, dark and grey solitons. While the bright soliton is a Gaussian-like intensity distribution with peak intensity, the dark soliton is a dark notch propagating on a uniform background intensity. A grey soliton can be thought of as a general case of a dark soliton where the dark notch is not entirely "dark", i.e., the minimum intensity is not zero but a small finite number. Figures 1.3 and 1.4 illustrate the dynamical evolution of bright and dark solitons.

Now, to understand optical spatial solitons in photorefractives, we need to understand a brief historical overview. The very first experimental observations of self trapping are described in [25] and utilized an experimental apparatus available in most modestly equipped laboratories. These were revolutionary results as there was no hint of self focusing in any previous experiments in related media, neither any theoretical understanding of the same. These experiments showed an unequivocal

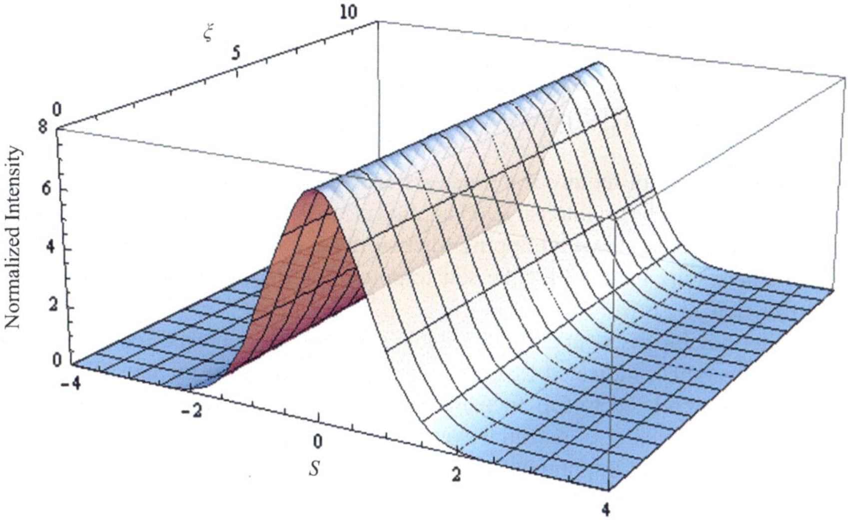

Fig. 1.3 Stable dynamical evolution of a bright soliton

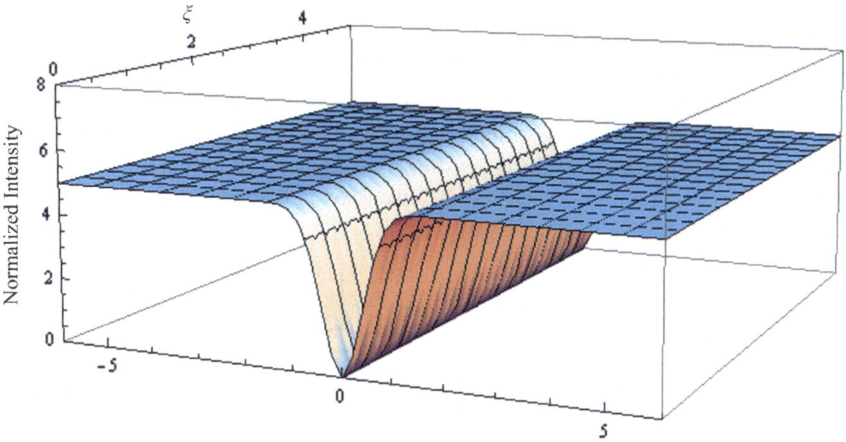

Fig. 1.4 Stable dynamical evolution of a dark soliton

self trapping of a light beam of visible wavelengths propagating in a biased photorefractive sample. The consequent spatial soliton was readily amenable to observation. It was also established that narrow light beams launched in biased photorefractive crystal would be able self-trap and in turn propagate robustly. Importantly, these beams remained self trapped and undistorted by beam fanning and sources of noise in the crystal. It was found that a 457 nm continuous wave of FWHM 15 μm could self trap without experiencing beam fanning for an external bias of 500 V/cm. Large deviances from optimal launch conditions also failed to distort these self trapped beams [26]. The initial observation of self trapping in photorefractive media led to rapid advancements and experiments to further study photorefractive solitons and provide the bedrock of phenomenological understanding of self trapping in photorefractives [25, 27]. There were various questions which were posed by these observations. Firstly, these observations pointed towards a transient self trapping process which did not require rigorous existence conditions expected for self trapping. Secondly, the beam intensity and applied electric field could be varied without significantly changing the self trapping whereas this is counterintuitive because these two parameters are markers for the self trapping nonlinearity which must balance the diffraction.

Another notable discovery was the stable self trapping achieved for two dimensional beams which indicated that the self trapping was not Kerr like. The catastrophic collapse related to self focusing in Kerr like media is well known. These aspects are still a matter of research even today. The quasi-steady state or the transient nature of the self trapping was a definite drawback. The transient nature of the self trapping results from the effect of the charge accumulation screening field E_0. The induced electro-optic lens flattens or saturates as the charges continue to separate until E_0 is totally screened. For stable self trapping, a compensating mechanism is needed through which accumulated charge can be eliminated by homogeneously illuminating the photorefractive crystal, which results in increasing the effective dark conductivity [3]. Depending on the relative intensity of the beam w.r.t the background, a dynamic equilibrium results in a steady state lensing effect. In Refs. [28] and [29], the authors were able to observe this steady state self trapped stable soliton in photorefractive crystals. A simple envisaged experimental setup for observing steady state photorefractive solitons is shown in Fig. 1.5. These type of solitons are called as "screening solitons" and will be discussed in the next section.

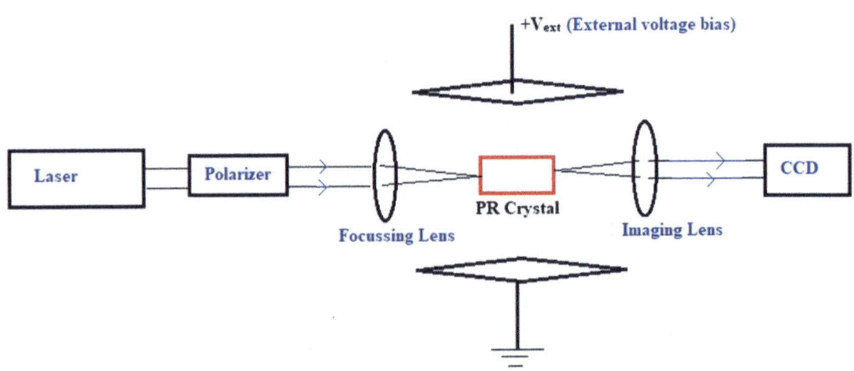

Fig. 1.5 Envisaged experimental setup for observing screening solitons [25, 27, 28, 30]

1.4 Theoretical Model for Photorefractive Screening Solitons

Screening solitons consist of the simplest type of optical spatial solitons in photorefractive media. They can be observed experimentally if the PR crystal is properly oriented by coinciding its principal axes with the *x*, *y*, *z* axes, and additionally has been externally biased. The drift current dominates in such a case and the space charge field formed can induce an index waveguide through the Pockels effect. This index waveguide then counteracts the effects of diffraction forming a stable spatial soliton. The external electric field screens the initial space charge field and hence the name "screening solitons".

Consider an optical beam that propagates in the *z*-direction permitted to diffract only in the x direction. For understanding a general theoretical foundation, let us consider that the crystal can be either noncentrosymmetric (SBN) or centrosymmetric (KLTN) or novel photorefractive crystals having both the linear and quadratic electro-optic effect (PMN-0.33PT) for illustrative purposes. The crystal's *c*-axis should be oriented along the *x*-direction. Additionally, assume that the beam is linearly polarized along the *x* direction and the external electric field is also applied in this *x*-direction. The refractive index along the *c*-axis $n_e^{'}$ is then given by [31, 32],

$$n_e^{'2} = n_e^2 - a n_e^4 r_{33} E_{sc} - b n_e^4 g_{eff} \epsilon_0^2 (\epsilon_r - 1)^2 E_{sc}^2 \tag{1.19}$$

where r_{33} is the electro-optic coefficient, n_e is the unperturbed refractive index along the extraordinary axis (or c-axis). $\vec{E}_{sc} = E_{sc}\hat{x}$ is the space charge field induced in the photorefractive crystal. The value of *a* and *b* is one or zero depending upon whether the material exhibits the linear electro-optic effect or the quadratic electro-optic effect or both. The electric field, as usual satisfies the Helmholtz equation,

$$\nabla^2 \vec{E} + (k_0 n_e^{'})^2 \vec{E} = 0 \tag{1.20}$$

where $k_0 = \frac{2\pi}{\lambda_0}$ is the wavenumber and λ_0 is the free space wavelength of the light beam.

Expressing $\vec{E} = \hat{x}\phi(x, z)\exp(ikz)$, and applying the slowly varying approximation to the envelope ϕ, one obtains the following paraxial equation,

$$i\phi_z + \frac{1}{2k}\phi_{xx} - \frac{k_0}{2}\left(n_e^3 r_{33} E_{sc}\right)\phi = 0 \tag{1.21}$$

where $k = k_0 n_e$, $\phi_z = \frac{\partial \phi}{\partial z}$ etc.

The Kukhtarev-Vinetskii [33] model is most commonly used to describe the charge transport and derive a relation for the space charge field. Under steady state conditions,

$$\gamma_R n N_D^+ = s_i(I + I_d)\left(N_D - N_D^+\right) \tag{1.22}$$

$$J = e\mu\left(nE_sc + \frac{kT}{e}\frac{\partial n}{\partial x}\right) \tag{1.23}$$

$$\frac{\partial J}{\partial x} = 0 \tag{1.24}$$

$$\frac{\partial E_{sc}}{\partial x} = \frac{e}{\epsilon_0\epsilon_r}\left(N_D^+ - N_A - n\right) \tag{1.25}$$

s_i is the photoionization cross section, J is the current density, γ_R is the carrier recombination rate, n is the free electron density, μ is the electron mobility, k_B is the Boltzmann constant, T is the absolute temperature, ϵ_r is the static relative permittivity. N_A is the acceptor or trap density, N_D is the donor concentration, $N_D{}^+$ is the ionized donor density. The thermally generated electrons also have a finite contribution to the space charge field and hence, to account for this, a quantity I_d is defined as the dark irradiance. $I = I(x,z)$ is the intensity profile of the optical beam. From the Poynting's theorem, we can express $I = (n_e/2\eta_0)|\phi|^2$ where $\eta_0 = (\mu_0/\epsilon_0)^{1/2}$. The fact that variables vary much more rapidly with x means that any spatial dependence on the z-direction can be neglected. In essence, (1.22)-(1.25) can be simultaneously solved to obtain the induced space charge field. In practice, this is quite difficult and certain approximations need to be done to simplify the ensuing analysis. In characteristic photorefractive media, N_D, $N_A >> n$ and $N_D^+ >> n$. So, (1.22)-(1.25) yield,

$$N_D^+ = N_A\left(1 + \frac{\epsilon_0\epsilon_r}{eN_A}\frac{\partial E_{sc}}{\partial x}\right) \tag{1.26}$$

$$n = \frac{s_i(N_D - N_A)}{\gamma_R N_A}(I + I_d)\left(1 + \frac{\epsilon_0\epsilon_r}{eN_A}\frac{\partial E_{sc}}{\partial x}\right)^{-1} \tag{1.27}$$

If the intensity I of the beam attains an asymptotically constant value at $x \to \pm\infty$, i.e., $I(x \to \pm\infty, z) = I_\infty$. Hence, this is a region of constant illumination and the space charge field has to be independent of x, i.e., $E_{sc}(x \to \pm\infty, z) = E_0$. If the spatial width of the optical beam is much less than the crystal width, then under a constant bias V, $E_0 \approx V/W$. From (1.27), the free electron density in the regions $x \to \pm\infty$ can be written,

$$n_0 = \frac{s_i(N_D - N_A)}{\gamma_R N_A}(I_\infty + I_d) \tag{1.28}$$

Also, as current density J is constant everywhere, so

$$n_0E_0 = nE_{sc} + \left(\frac{k_BT}{e}\right)\left(\frac{\partial n}{\partial x}\right) \tag{1.29}$$

from which we can get,

$$E_{sc} = \frac{n_0 E_0}{n} - \frac{k_B T}{e}\frac{1}{n}\frac{\partial n}{\partial x} \tag{1.30}$$

Substitution of (1.28) into (1.30) gives,

$$E_{sc} = E_0 \frac{I_\infty + I_d}{I + I_d}\left(1 + \frac{\epsilon_0 \epsilon_r}{eN_A}\frac{\partial E_{sc}}{\partial x}\right) - \frac{k_B T}{e}\frac{(\partial I/\partial x)}{(I + I_d)}$$
$$+ \frac{k_B T}{e}\frac{\epsilon_0 \epsilon_r}{eN_A}\left(1 + \frac{\epsilon_0 \epsilon_r}{eN_A}\frac{\partial E_{sc}}{\partial x}\right)^{-1}\frac{\partial^2 E_{sc}}{\partial x^2} \tag{1.31}$$

If the external bias is strong, the value of E_0 will reach appreciable limits and the drift component of the current will be dominant. Hence, all the terms related to the diffusion, i.e., the terms with $k_B T/e$ dependence can be neglected as they can be considered small as compared to the drift component. In addition, if the intensity $I(x, z)$ of the optical beam varies slowly with respect to x, then the term $\frac{\epsilon_0 \epsilon_r}{eN_A}\frac{\partial E_{sc}}{\partial x}$ will be much less than unity in typical PR media. Thus, the space charge field can be approximated to be,

$$E_{sc} = E_0 \frac{I_\infty + I_d}{I + I_d} \tag{1.32}$$

Inserting (1.32) into (1.20), one can establish the soliton envelope evolution equation. Using the following dimensionless co-ordinates,$\xi = z/\left(kx_0^2\right)$, $s = x/x_0$, $\phi = (2\eta_0 I_d/n_e)^{1/2} U$, U can be shown to satisfy the following equation,

$$iU_\xi + \frac{1}{2}U_{ss} - \frac{\beta_1(1+\rho)U}{1+|U|^2} - \frac{\beta_2(1+\rho)^2 U}{\left(1+|U|^2\right)^2} = 0 \tag{1.33}$$

where x_0 is an arbitrary scale parameter and intensity has been scaled with respect to the dark irradiance as $I = I_d|U|^2$. Also, $\rho = I_\infty/I_d$ and $\beta_1 = a(k_0 x_0)^2\left(n_e^4 r_{33}/2\right)E_0$ and $\beta_2 = b(k_0 x_0)^2\left(n_e^4 g_{eff}\epsilon_0\left(\epsilon_r - 1\right)^2/2\right)E_0^2$.

1.4.1 Bright Soliton States

For examining the bright spatial soliton states, let us assume a SBN crystal for which we can observe only the Pockels' or the linear electro-optic effect. Hence $a = 1$, $b = 0$. The optical beam intensity will vanish at $s \to \infty$ and therefore $I(\infty) = \rho = 0$. So Eq. (1.15) becomes,

$$iU_\xi + \frac{1}{2}U_{ss} - \frac{\beta_1 U}{1+|U|^2} = 0 \tag{1.34}$$

Expressing the beam envelope as $U = r^{1/2}y(s)\exp(i\nu\xi)$ where ν represents a nonlinear shift of the propagation constant and y(s) is a bounded function where $0 \leq y(s) \leq 1$. For bright solitons, the following boundary conditions need to be satisfied,

$$y(0) = 1,\ \dot{y}(0) = 0,\ y(s \to \pm\infty) = 0,\ \dot{y}(s \to \pm\infty) = 0.$$

Putting U into (1.34) leads to,

$$\ddot{y} - 2\nu y - 2\beta_1 \frac{y}{1 + ry^2} = 0 \tag{1.35}$$

where $\ddot{y} = \frac{dy}{ds}$.

By integrating (1.35) once, and using the boundary conditions for the bright solitons as enumerated above,

$$\dot{y}^2 = \frac{2\beta_1}{r}\left[-y^2 \ln(1 + r) + \ln\left(1 + ry^2\right)\right] \tag{1.36}$$

$$\nu = -\frac{\beta_1}{r} \ln(1 + r) \tag{1.37}$$

Integrating once more,

$$(2\beta_1)^{1/2} s = \pm \int_y^1 \frac{r^{1/2} d\hat{y}}{\left[ln\left(1 + r\hat{y}^2\right) - \hat{y}^2 ln(1 + r)\right]^{1/2}} \tag{1.38}$$

Equation (1.38) does not have any closed form solution so simple numerical integration can be used to find out the bright field profile y(s).

The RHS of (1.38) will be positive when $\beta_1 > 0$ and hence $E_0 > 0$ is the condition for formation of a bright soliton. Considering the following parameters of the SBN crystal [31], $n_e = 2.35$, $r_{33} = 224 \times 10^{-12}$ m/V and $\lambda_0 = 0.5\mu m$, $x_0 = 40\mu m$, $E_0 = 1.9 \times 10^5$ V/m. Taking these into account, one can obtain, $\beta_1 \approx 164.36$. Figure 1.6 shows the normalized intensity profile of the bright soliton for $r = 1$. Figure 1.7 shows the existence curve for the bright soliton, i.e., the variation of FWHM of the soliton intensity profile with r.

1.4.2 Dark and Grey Soliton States

Dark soliton beams have an antisymmetric field profile with respect to x, the spatial co-ordinate. They embedded on a constant intensity background so I_∞, ρ are finite. Again, considering the case of a photorefractive crystal exhibiting the linear electro-optic effect, the evolution eqn is,

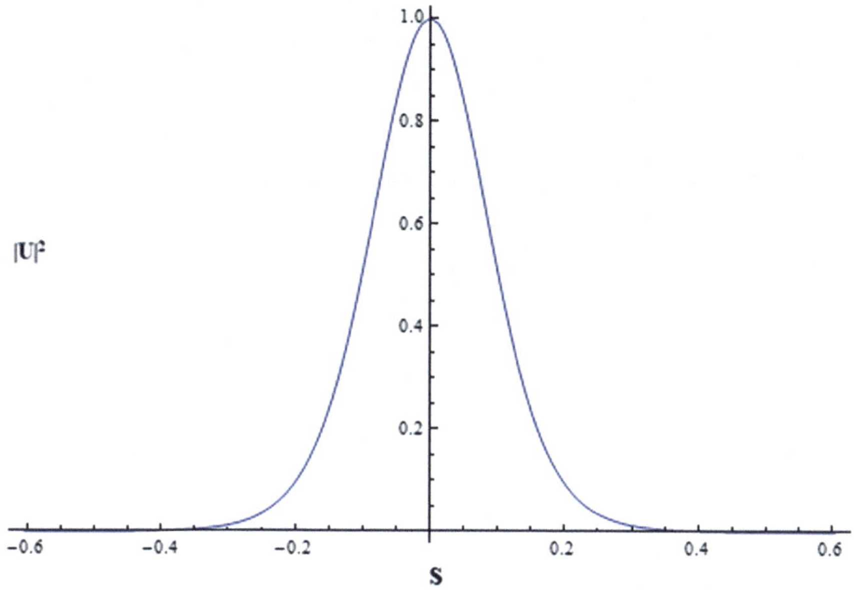

Fig. 1.6 Normalized intensity profile of the bright soliton for $r = 1$; $\beta_1 \approx 164.36$ [31]

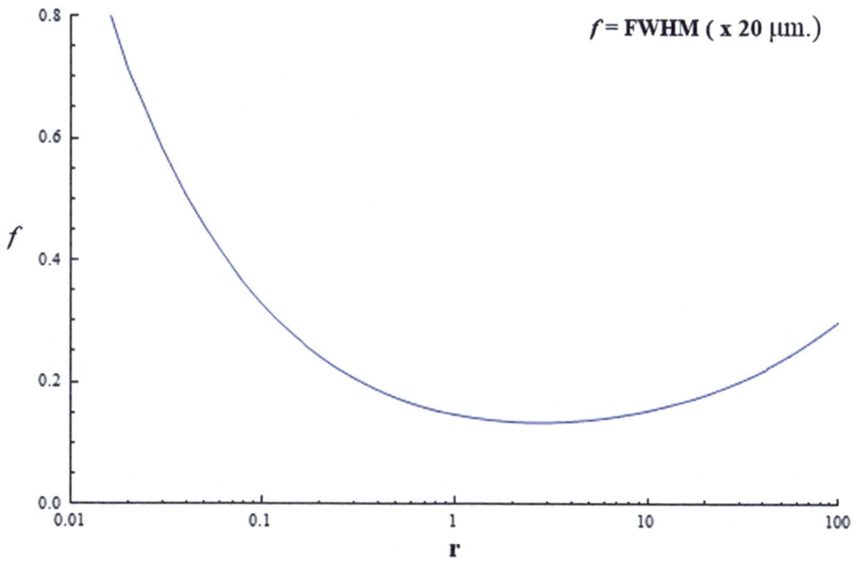

Fig. 1.7 Existence curve for the bright soliton [31]

$$iU_\xi + \frac{1}{2}U_{ss} - \frac{\beta_1(1+\rho)U}{1+|U|^2} = 0 \tag{1.39}$$

Let $U = \rho^{1/2}y(s)\exp(i\nu\xi)$ where ν represents a nonlinear shift of the propagation constant and y(s) is a bounded function where $0 \le y(s) \le 1$. For dark solitons, the following boundary conditions need to be satisfied,

$$y(0) = 0,\ \dot{y}(0) = 0,\ y(s \to \pm\infty) = 1,\ \dot{y}(s \to \pm\infty) = 0.$$

Substitution of the aforementioned ansatz of U into (1.39) leads to,

$$\ddot{y} - 2\nu y - 2\beta_1(1+\rho)\frac{y}{1+\rho y^2} = 0 \tag{1.40}$$

where $\ddot{y} = \frac{dy}{ds}$.

By using the boundary conditions for the dark solitons as enumerated above and integrating (1.40) once, we get,

$$\dot{y}^2 = (-2\beta_1)\left[\left(y^2-1\right) - \frac{1+\rho}{\rho}\ln\left(\frac{1+\rho y^2}{1+\rho}\right)\right] \tag{1.41}$$

$$\nu = -\beta_1 \tag{1.42}$$

Equation (1.41) does not have any closed form solution so simple numerical integration can be used to find out the dark field profile y(s). The RHS of (1.41) will be positive when $\beta < 0$ and hence $E_0 < 0$ is the condition for formation of a dark soliton. This means that reversing the polarity of the external bias field can render the same photorefractive crystal (SBN in this case) to support a dark screening soliton. So we take $E_0 = -1.8 \times 10^5$ V/m and all other parameters are taken same as in the previous case. Figure 1.8 shows the normalized intensity profile for the dark soliton. Figure 1.9 shows the existence curve for the dark soliton.

There is another class of interesting soliton solutions, known as grey solitons. This family is also expected to evolve according to (1.39) as they are also embedded on a constant intensity background. The grey solitons can be thought of as a general case of dark solitons where the dark notch is not completely "dark". The envelope U is expressed as,

$$U = \rho^{1/2}y(s)\exp\left[i\left(\nu\xi + \int^s \frac{J ds'}{y^2(s')}\right)\right] \tag{1.43}$$

J is a real constant to be determined. Further, $y(s \to \pm\infty) = 1$ and $y^2(0) = m$, $\dot{y}(0) = 0$. As usual, all derivatives of y at infinity are zero.

Substituting (1.43) in (1.39) results in,

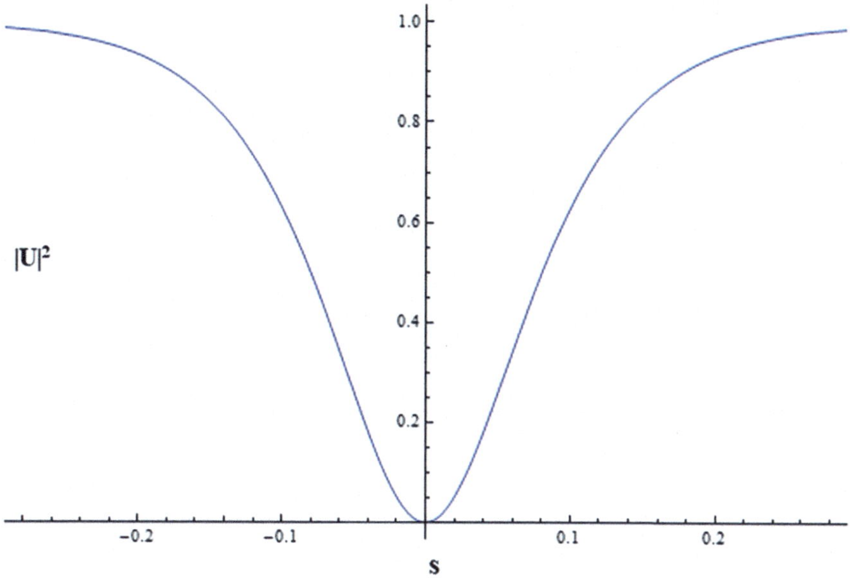

Fig. 1.8 Normalized intensity profile of the dark soliton for $\rho = 1$; $\beta_1 \approx -164.36$ [31]

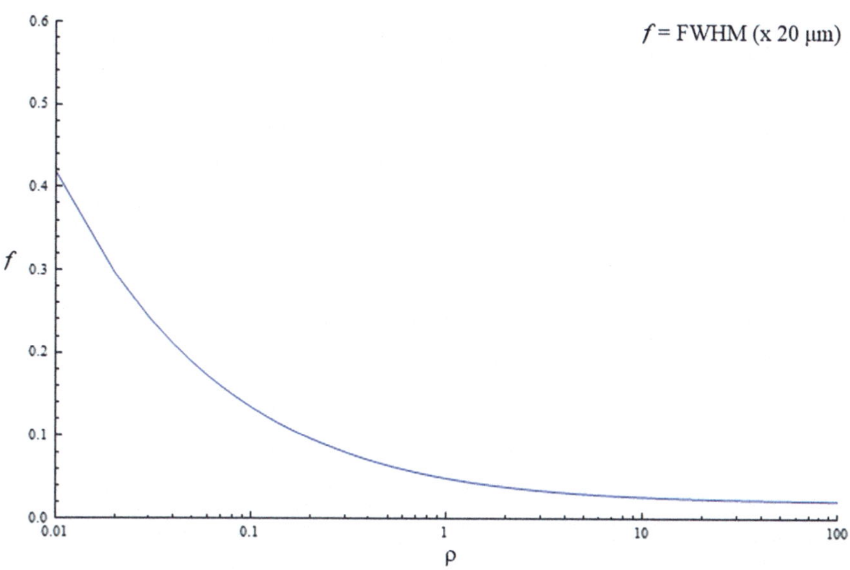

Fig. 1.9 Existence curve for the dark soliton [31]

$$\ddot{y} - 2\nu y - \frac{J^2}{y^3} - 2\beta_1(1+\rho)\frac{y}{1+\rho y^2} = 0 \tag{1.44}$$

Using the boundary conditions of y at infinity,

$$J^2 = -2(\nu + \beta_1) \tag{1.45}$$

Further integrating (1.45),

$$(\dot{y})^2 = 2\nu\left(y^2 - 1\right) + \frac{2\beta_1}{\rho}(1+\rho)\ln\left(\frac{1+\rho y^2}{1+\rho}\right) + 2(\nu+\beta_1)\left(\frac{1-y^2}{y^2}\right) \tag{1.46}$$

$$\nu = \frac{-\beta_1}{(m-1)^2}\left[\frac{m(1+\rho)}{\rho}\ln\left(\frac{1+\rho m}{1+\rho}\right) + (1-m)\right] \tag{1.47}$$

(1.47) can be solved numerically to find the envelope *y(s)*. It is notable that the parameters β_1, ρ, m have to be selected such that RHS of (1.46) is positive and J^2 is positive. The condition for existence of grey solitons can be found out to be $\beta_1 < 0$ and $m < 1$.

1.4.3 Self Deflection of Photorefractive Solitons

Till now, we have assumed that the diffusion effect is negligible if we use a relatively strong external field. It is now quite pertinent to look for the effect of the diffusion space charge field on the spatial soliton evolution. It has been observed that the diffusion effect results in an asymmetric tilt in the induced waveguide in the photorefractive crystal which, in turn results in significant changes in how the dynamical evolution of the soliton takes place. Assuming a slow variation of the intensity, i.e., we shall neglect the $\frac{\epsilon_0\epsilon_r}{eN_A}\frac{\partial E_{sc}}{\partial x}$ terms but not the $k_B T$ terms, the expression for the space charge field (1.31) becomes,

$$E_{sc} = E_0\frac{I_d}{I+I_d} - \frac{k_B T}{e}\frac{\partial I/\partial x}{I+I_d} \tag{1.48}$$

Of course, E_0 can be approximated to $\pm V/W$ if the spatial extent of the wave is less than the width of the photorefractive crystal. Also, as in the previous section, let us consider a photorefractive crystal exhibiting the linear electro-optic effect only. So, the dynamical evolution equation for the optical beam envelope becomes,

$$iU_\xi + \frac{1}{2}U_{ss} - \frac{\beta_1(1+\rho)U}{1+|U|^2} + \gamma\frac{\left(|U|^2\right)_s}{1+|U|^2} = 0 \tag{1.49}$$

where we have used the dimensionless co-ordinates and the symbols have their usual meaning as defined before. Now, as we know, the diffusion effect is very small if the external bias is relatively large and the drift component of the current dominates over the diffusion component. Hence, we can treat the diffusion as a perturbation to the dynamical evolution equation with $\gamma = 0$. So, we shall first numerically find out the solution for *y(s)* satisfying (1.49) with $\gamma = 0$,

$$(2\beta_1)^{1/2}s = \pm \int_y^1 \frac{r^{1/2}d\hat{y}}{[ln(1+r\hat{y}^2) - \hat{y}^2 ln(1+r)]^{1/2}} \tag{1.50}$$

Taking this solution for the soliton, now substituting this into (1.49) with finite γ, we can find the dynamical evolution of the soliton beam in the longitudinal direction. Given the value of *y(s)* at the beginning, $\xi = 0$, we need to find *y(s)* at further values of ξ. For this, a simple finite difference formulation will suffice and is known as the finite difference beam propagation method [34]. There is another beam propagation method based on Fourier transforms, known as the split step fourier method, for which the reader is referred to [35].

Assuming the typical SBN crystal's parameters are used [31], the dynamical evolution of the bright soliton is shown in Fig. 1.10. We can infer a few things here. Firstly, the center of the soliton beam follows a parabolic trajectory. Secondly, the soliton's intensity profile remains invariant when propagating implying an adiabatic evolution.

In addition to the above approach, there are perturbative mathematical methods to investigate the effect of the diffusion space charge field. Such methods have been widely used in nonlinear fiber optics as well as photorefractive nonlinear optics. Since, we can see an approximately adiabatic evolution of the beam, we start with

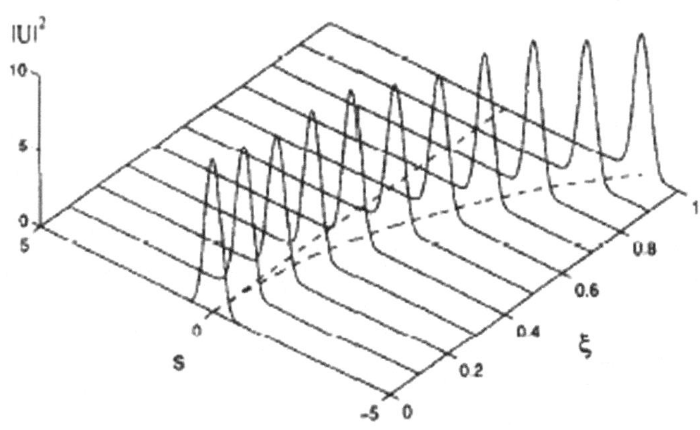

Fig. 1.10 Evolution of the intensity profile of the soliton, $r = 10$, $\beta = 34.5$, $\gamma = 0.56$. (Reprinted from M.I. Carvalho, S.R. Singh, D.N. Christodoulides, Self-deflection of steady-state bright spatial solitons in biased photorefractive crystals. Opt. Commun. **120**, 311–315. Copyright 1995, with permission from Elsevier)

the following ansatz for the soliton solution,

$$U(\xi, s) = r^{1/2}y[s + v(\xi)] \times \exp(i\{\mu\xi + \omega(\xi)[s + v(\xi)] + \alpha(\xi)\}) \tag{1.51}$$

$v(\xi)$ is the shift in the position of the beam center, $\omega(\xi)$ is the angle between the central wave vector and the propagation axis, and $\alpha(\xi)$ is the variable phase of the beam. For obtaining the equations of motion of these variables, substitute (1.51) into the two complex conservation laws of (1.49). These complex conservation laws can be easily found out by multiplying (1.49) by U and U^* and iU_s and $iU_s{}^*$ and then integrating over s. This yields,

$$dv/d\xi = \omega \tag{1.52}$$

$$d\alpha/d\xi = \omega^2/2 \tag{1.53}$$

$$d\omega/d\xi = 4\beta\gamma K(r) \tag{1.54}$$

where,

$$K(r) = \int_{-\infty}^{\infty} ds \frac{2ry^2}{1 + ry^2(s)} \times \left\{y^2(s)\ln(1 + r) - \ln\left[1 + ry^2(s)\right]\right\} \times \left(\int_{-\infty}^{\infty} dsry^2(s)\right)^{-1} \tag{1.55}$$

Since we cannot solve analytically for *y(s)*, *K(r)* has to be found numerically. Figure 1.11 shows the variation of K with r.

Equations (1.52)–(1.54) can be integrated to give,

$$\omega(\xi) = 4\beta\gamma K(r)\xi \tag{1.56}$$

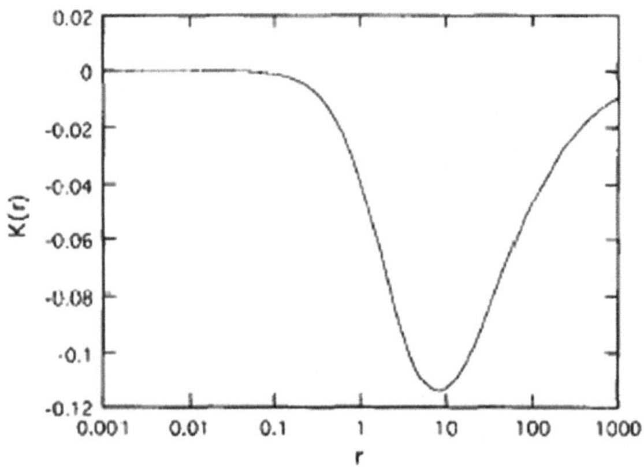

Fig. 1.11 Dependence of K function on *r*. (Reprinted from M.I. Carvalho, S.R. Singh, D.N. Christodoulides, Self-deflection of steady-state bright spatial solitons in biased photorefractive crystals. Opt. Commun. **120**, 311–315. Copyright 1995, with permission from Elsevier)

$$v(\xi) = -2\beta\gamma K(r)\xi^2 \tag{1.57}$$

$$\alpha(\xi) = 8[\beta\gamma K(r)]^2\xi^3/3 \tag{1.58}$$

We can see clearly that in case of no diffusion, i.e., $\gamma = 0$, $\omega = \alpha = v = 0$. In contrast, if diffusion is taken into account, (1.57) shows a parabolic trajectory for the beam center. (1.56) shows that the central frequency shifts linearly with the propagation distance. From these relations, we can immediately infer that the beam has undergone a displacement of, $x_d = (n_e^3 r_{33} k_0)^2 (k_B T/2e) E_0 K(r) z^2$ where z is the actual distance moved.

1.5 Concluding Remarks and Further Reading

We have tried to present a succinct but clear introduction of the photorefractive solitons in the present chapter. The electro-optic effect has been explained with a relevant example which then leads to the explanation of the photorefractive effect and self trapping in photorefractive media. A brief experimental overview has also been given. Finally, we have discussed a comprehensive theory for the simplest type of photorefractive solitons, known as screening solitons. The derivation of the induced space charge field is discussed taking into consideration the Kukhtarev charge transport model. The normalized intensity profiles are obtained alongwith the soliton width existence curves for all three types of solitos, viz. bright, dark and grey. The dynamical evolution of these solitons is then discussed and the effect of the diffusion on the propagation of the soliton is investigated.

It has been our intention to acquaint the reader with the basic concepts needed to proceed onward for an understanding of photorefractive solitons. The photorefractive effect has been around for much longer and many nonlinear optical phenomena other than solitons, like the holographic storage, wave mixing, phase conjugation, ring resonators etc. have been studied in photorefractive materials. If the reader wishes to have an in depth understanding of the photorefractive effect and different photorefractive materials, Refs. [1, 2, 36, 37] would be a good place to start for a classical foundation. References [3, 4] provide a fresh perspective with many newer discoveries in photorefractives explained and summarized. Reference [38] is a review article also worth reading for the same reason.

References

1. J. Frejlich, *Photorefractive Materials: Fundamental Concepts, Holographic Recording and Materials Characterization* (Wiley, 2007)
2. C. Dainty, *Introduction to Photorefractive Nonlinear Optics*, vol. 27, no. 2. (Wiley, 1995)

3. P. Günter, J. Huignard, A. Glass, *Photorefractive Materials and Their Applications 1* (Springer Science+Business Media Inc., New York, 2006).
4. P. Günter, J.P. Huignard, *Photorefractive Materials and Their Applications 2*, vol. 114 (Springer Science and Business Media, New York, 2007).
5. S. Trillo, W.E. Torruellas (eds.), *Spatial Solitons, Springer Series in Optical Sciences*, vol. 31 (Springer, Berlin, 2001)
6. D.D. Nolte, *Photorefractive Effects and Materials* (Springer Science & Business Media, New York, 2013).
7. A. Ashkin et al., Optically-induced refractive index inhomogeneities in $LiNbO_3$, and $LiTaO_3$. in *Landmark Papers on Photorefractive Nonlinear Optics* (World scientific, 1995), pp. 29–31.
8. M. Aguilar, M. Carrascosa, Photorefractive effect and nonlinear susceptibilities. Elsevier **5**(March), 187–192 (1996)
9. F. Cooper, H. Shepard, Solitons in the Camassa-Holm shallow water equation. Phys. Lett. A **194**(4), 246–250 (1994). https://doi.org/10.1016/0375-9601(94)91246-7
10. N. Zabusky, C. Galvin, Shallow-water waves, the Korteweg-de Vries equation and solitons. J. Fluid Mech. **47**, 811–824 (1971)
11. A.I. Maimistov, Solitons in nonlinear optics. Quantum Electron. **40**(9), 756–781 (2010). https://doi.org/10.1070/QE2010v040n09ABEH014396
12. E.A. Kuznetsov, Solitons in a parametrically unstable plasma. Doklady Akademii Nauk SSSR **236**(JANUARY), 575–577 (1977)
13. G. Lenz, P. Meystre, E.M. Wright, Nonlinear atom optics: General formalism and atomic solitons. Phys. Rev. A **50**(2), 1681–1691 (1994). https://doi.org/10.1103/PhysRevA.50.1681
14. D. Neshev, E. Ostrovskaya, Y. Kivshar, W. Krolikowski, Spatial solitons in optically induced gratings. Opt. Lett. **28**(9), 710 (2003). https://doi.org/10.1364/OL.28.000710
15. S. Burger et al., Dark solitons in Bose-Einstein condensates. Phys. Rev. Lett. **83**(25), 5198–5201 (1999). https://doi.org/10.1103/PhysRevLett.83.5198
16. G.I. Stegeman, Optical spatial solitons and their interactions: universality and diversity. Science **286**(5444), 1518–1523 (1999). https://doi.org/10.1126/science.286.5444.1518
17. J.S. Russell, Report on waves. in *14th Meeting of the British Association for the Advancement of Science* (1844), pp. 311–390
18. C.S. Gardner, J.M. Greene, M.D. Kruskal, R.M. Miura, Method for solving the Korteweg-deVries equation. Phys. Rev. Lett. **19**(19), 1095–1097 (1967). https://doi.org/10.1103/PhysRevLett.19.1095
19. C.S. Gardner, J.M. Greene, M.D. Kruskal, R.M. Miura, Korteweg-devries equation and generalizations. VI methods for exact solution. Commun. Pure Appl. Math. **27**(1), 97–133 (1974). https://doi.org/10.1002/cpa.3160270108
20. P. Butcher, D. Cotter, *The Elements of Nonlinear Optics*, vol. 9 (Cambridge University Press, 1991)
21. Y.R. Shen, *The principles of Nonlinear Optics* (Wiley-Interscience, 1984)
22. R.W. Boyd, *Nonlinear Optics* (Academic, San Diego, 1992).
23. F. Reynaud, A. Barthelemy, Optically controlled interaction between two fundamental soliton beams. Europhys. Lett. (EPL) **12**(5), 401–405 (1990). https://doi.org/10.1209/0295-5075/12/5/004
24. Y.S. Kivshar, G.P. Agrawal, *Optical Solitons: From Fibers to Photonic Crystals* (2003)
25. G.C. Duree et al., Observation of self-trapping of an optical beam due to the photorefractive effect. Phys. Rev. Lett. **71**(4), 533–536 (1993). https://doi.org/10.1103/PhysRevLett.71.533
26. M. Segev et al., Stability of photorefractive spatial solitons. Opt. Lett. **19**(17), 1296–1298 (1994)
27. B. Crosignani, M. Segev, D. Engin, P. Di Porto, A. Yariv, G. Salamo, Self-trapping of optical beams in photorefractive media. JOSA B **10**(3), 446–453 (1993)
28. B. Crosignani, G. Salamo, G.C. Valley, M. Segev, P. Di Porto, M.-F. Shih, Observation of two-dimensional steady-state photorefractive screening solitons. Electron. Lett. **31**(10), 826–827 (1995). https://doi.org/10.1049/el:19950570

29. K. Kos, H. Meng, G. Salamo, M. Shih, M. Segev, G.C. Valley, One-dimensional steady-state photorefractive screening solitons. Phys. Rev. E Stat. Phys. Plasmas Fluids Relat. Interdiscip. Topics **53**(5), R4330–R4333 (1996). https://doi.org/10.1103/PhysRevE.53.R4330
30. M. Segev, B. Crosignani, A. Yariv, B. Fischer, Spatial solitons in photorefractive media. Phys. Rev. Lett. **68**(7), 923–926 (1992). https://doi.org/10.1103/PhysRevLett.68.923
31. D.N. Christodoulides, M.I. Carvalho, Bright, dark, and gray spatial soliton states in photorefractive media. J. Opt. Soc. Am. B **12**(9), 1628 (1995). https://doi.org/10.1364/JOSAB.12.001628
32. M. Segev, A. J. Agranat. Spatial solitons in centrosymmetric photorefractive media. Opt. lett. **22**(17) 1299–1301 (1997).
33. N.V. Kukhtarev, V.B. Markov, S.G. Odulov, M.S. Soskin, V.L. Vinetskii, Holographic storage in electrooptic crystals. i. steady state. Ferroelectrics **22**(1), 949–960 (1978). https://doi.org/10.1080/00150197908239450
34. Y. Chung, N. Dagli, An assessment of finite difference beam propagation method. IEEE J. Quantum Electron. **26**(8), 1335–1339 (1990)
35. J.A.C. Weideman, B.M. Herbst, Split-step methods for the solution of the nonlinear Schrödinger equation. SIAM J. Numer. Anal. **23**(3), 485–507 (1986)
36. P. Yeh, C. Gu, *Landmark Papers on Photorefractive Nonlinear Optics* (World Scientific, 1995)
37. D.D. Nolte, *Photorefractive Effects and Materials*, vol. 27 (1995)
38. Z. Chen, M. Segev, D.N. Christodoulides, Optical spatial solitons: historical overview and recent advances. Rep. Prog. Phys. **75**(8), 086401 (2012). https://doi.org/10.1088/0034-4885/75/8/086401

Chapter 2
Photovoltaic and Pyroelectric Solitons

2.1 The Photovoltaic Effect

The conventional photovoltaic effect is a phenomenon which describes the generation of a voltage or electric current in a photovoltaic cell when sunlight is incident upon it. The cells within a solar panel convert sunlight to electrical energy. The solar cells are made of two joined *p*-type and *n*-type semiconductors which result in a p–n junction. A space charge field is formed at the junction as electrons diffuse to the p-side and holes diffuse to the *n*-side. This field causes a potential difference which can be harnessed as electrical energy.

The photovoltaic effect in which we are interested is different from the above described p–n junction photovoltaic effect described in solar cells. The bulk photovoltaic effect is known to occur in semiconductors and insulators. The bulk photovoltaic effect is also known to as "anomalous". That is so because the typical photovoltage produced by incident light is much greater than the band gap of the semiconductor. In certain crystals, the photovoltage may be of the order of $\sim 10^3$ V.

The main phenomenon behind the bulk photovoltaic effect is that various electron related processes occur with different rates in different directions. Photo-excitation, scattering, and relaxation have a different probability of occuring in different directions with respect to the motion of the electron. This results in generation of a large photovoltage [1, 2].

Another mechanism involves development of parallel stripes ferroelectric domains in certain materials. Each domain acts like a photovoltaic and the domain wall behaves like a contact connecting the adjacent photovoltaics. The domains add in series, and hence the overall open-circuit voltage is quite large.

In Fig. 2.1, a simple system is illustrated through which we can understand the bulk photovoltaic effect. Consider two electronic levels separated by an energy gap of say, 3 eV in a unit cell. The blue and pink arrows show radiative and non radiative transitions respectively. An electron can move from A to B by absorbing a photon. Conversely, it may move from B to A by emitting a photon. The purple arrows indicate non radiative transitions. Here, it is implied that an electron can move from

A. Katti and R. A. Yadav, *Optical Spatial Solitons in Photorefractive Materials*, Progress in Optical Science and Photonics 14, https://doi.org/10.1007/978-981-16-2550-3_2

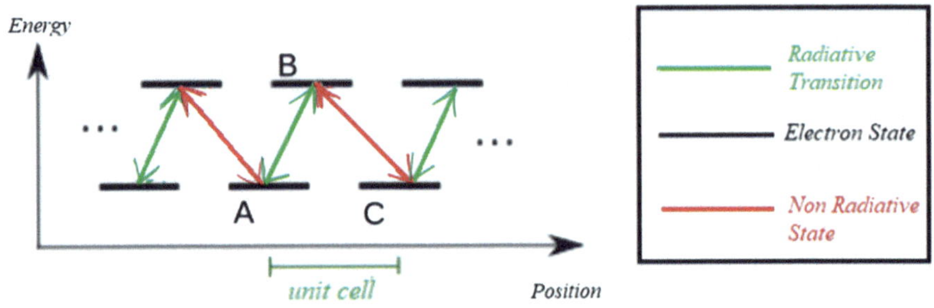

Fig. 2.1 The photovoltaic effect illustrated

B to C via lattice vibrations or emitting phonons, or vice versa by absorption of phonons.

If a light beam is incident upon a photovoltaic crystal, considering the above scenario, an electron can go from A to B to C by absorbing photons. But, the electron does not move in the opposite direction, i.e., from C to A through B, since the shift from C to B proceeds if a large thermal variation is present which in turn is improbable. Hence, we see a net rightward photocurrent.

There are many interesting features of the bulk photovoltaic effect distinguishing it from the conventional photovoltaic effect. Within the region of power generation in the characteristic I-V curve, electrons and holes are move towards higher and lower fermi levels respectively. We expect the opposite based on the drift diffusion equation. For example, power generation in a silicon solar cell is possible due to splitting of the quasi-fermi levels which implies the fact that the motion of electrons is towards the decreasing quasi Fermi level and the motion of holes is towards the increasing quasi Fermi level as per the drift diffusion equation. In contrast, power is generated in a bulk photovoltaic without any splitting of quasi-fermi levels.

The drift diffusion effect predicts that freely moving electrons will lessen the photocurrent and consequently diminish the photovoltaic effect. So, appreciable open-circuit voltages are observed only in crystals that exhibit very low dark conductivity.

The net motion of electrons due to the bulk photovoltaic effect is in the opposite direction to that expected due to the drift–diffusion equation. Hence, the quantum efficiency lessens considerably even for a thick device. Large amount of photons (of the order of 10^6) may be needed to transport an electron between the two electrodes. An increase in thickness results in voltage going up and a decrease in current. Also, the current may have different directions depending on the light polarization. Such effects are unheard of in silicon or any other ordinary solar cell.

2.2 Screening Photovoltaic Solitons

2.2.1 Theoretical Foundation

Until now, we have considered an externally biased non photovoltaic photorefractive crystal for studying steady state optical spatial solitons. However, if we take a photorefractive crystal also having a finite photovoltaic coefficient, the question arises as to whether it can support optical spatial solitons in the presence of an external bias? In [3], the authors have studied this in detail and we shall study these screening photovoltaic solitons in this section. Taking the electric field envelope as $\overrightarrow{E} = \hat{x}\phi(x, z)\exp(ikz)$, the paraxial equation of diffraction is [4],

$$i\frac{\partial \phi}{\partial z} + \frac{1}{2k}\frac{\partial^2 \phi}{\partial x^2} - \frac{k_0 n_e^3 r_{eff} E_{sc}}{2}\phi = 0 \tag{2.1}$$

The induced space charge field can be derived from the set of rate equations, continuity equations and Gauss law in one dimension for steady state [5],

$$\gamma_R n N_D^+ = s_i(I + I_d)\left(N_D - N_D^+\right) \tag{2.2}$$

$$\frac{\partial E_{sc}}{\partial x} = \frac{e}{\epsilon_0 \epsilon}\left(N_D^+ - N_A - n\right) \tag{2.3}$$

$$J = e\mu n E_{sc} + k_B T \mu \frac{\partial n}{\partial x} + k_p s_i\left(N_D - N_D^+\right)I \tag{2.4}$$

$$\frac{\partial J}{\partial x} = 0 \tag{2.5}$$

where the symbols have their usual meanings. There is one change in these equations if we compare them with those used in previous chapters. We have now considered k_{p} to be the photovoltaic constant which contributes the photovoltaic current term. As usual, any z spatial dependence has been ignored assuming a much more rapid variation in x. Now, in typical photovoltaic-photorefractive media, $N_D^+ \gg n$, $N_D \gg n$, and $N_A \gg n$. Hence, (2.2) and (2.3) give,

$$N_D^+ = N_A\left(1 + \frac{\epsilon_0 \epsilon_r}{e N_A}\frac{\partial E_{sc}}{\partial x}\right) \tag{2.6}$$

$$n = \frac{s_i(N_D - N_A)}{\gamma_R N_A}(I + I_d)\left(1 + \frac{\epsilon_0 \epsilon_r}{e N_A}\frac{\partial E_{sc}}{\partial x}\right)^{-1} \tag{2.7}$$

If the intensity of the light beam varies relatively slowly with respect to x, the term $\frac{\epsilon_0 \epsilon_r}{e N_A}\frac{\partial E_{sc}}{\partial x}$ can be ignored, as it is of the order of much less than unity. So, from

(2.6) and (2.7) it can be inferred that,

$$N_D^+ = N_A \tag{2.8}$$

$$n = \frac{s_i(N_D - N_A)}{\gamma_R N_A}(I + I_d) \tag{2.9}$$

In regions of constant illumination, we know, $I(x \to \pm\infty, z) = I_\infty$, $E_{sc}(x \to \pm\infty, z) = E_0$ where E_0 is the external bias field. From (2.7), electron density at $x \to \pm\infty$, i.e., n_∞ can be obtained as,

$$n_\infty = \frac{s_i(N_D - N_A)}{\gamma_R N_A}(I_\infty + I_d) \tag{2.10}$$

From (2.4), we know that,

$$J_\infty = J(x \to \pm\infty, z) = e\mu n_\infty E_0 + k_p s_i (N_D - N_A) I_\infty \tag{2.11}$$

Substituting (2.10) into (2.11),

$$J_\infty = e\mu n_\infty \left(E_0 + E_p \frac{I_\infty}{I_\infty + I_d} \right) \tag{2.12}$$

where, $E_p = k_p \gamma_R N_A / (e\mu)$.

Again, from (2.2) and (2.4), we get,

$$J = e\mu n \left(E_{sc} + \frac{k_B T}{e} \frac{\partial \ln n}{\partial x} + E_p \frac{I}{I + I_d} \right) \tag{2.13}$$

From (2.5), we can infer that the current is constant everywhere, so, from (2.12) and (2.13),

$$n_\infty \left(E_0 + E_p \frac{I_\infty}{I_\infty + I_d} \right) = n \left(E_{sc} + \frac{k_B T}{e} \frac{\partial \ln n}{\partial x} + E_p \frac{I}{I + I_d} \right) \tag{2.14}$$

Finally, we can obtain the space charge field from (2.14) as,

$$E_{sc} = E_0 \frac{I_\infty + I_d}{I + I_d} + E_p \frac{I_\infty - I}{I + I_d} - \frac{k_B T}{e} \frac{1}{I + I_d} \frac{\partial I}{\partial x} \tag{2.15}$$

The final dynamical evolution equation can now be setup by substituting (2.15) into (2.1),

$$i\frac{\partial U}{\partial \xi}+\frac{1}{2}\frac{\partial^2 U}{\partial s^2}-\beta(\rho+1)\frac{1}{1+|U|^2}U-\alpha\frac{\left(\rho-|U|^2\right)}{1+|U|^2}U+\gamma\frac{\left(|U|^2\right)_s}{1+|U|^2}U=0 \tag{2.16}$$

where we use the dimensionless co-ordinates, $\xi = z/\left(kx_0^2\right), s = x/x_0, \phi = (2\eta_0 I_d/n_e)^{1/2}U$ with x_0 to be an arbitrary spatial width and the intensity scaled with the dark irradiance I_{d}, $\rho = I_\infty/I_d, \beta = (k_0x_0)^2\left(n_e^4 r_{eff}/2\right)E_0$, $\alpha = (k_0x_0)^2\left(n_e^4 r_{eff}/2\right)E_p, \gamma = \left(k_0^2 x_0 n_e^4 r_{eff}\right)k_B T/(2e)$.

2.2.2 *Spatial Soliton States*

Once the dynamical evolution (2.16) has been obtained, it is a simple matter now to solve this get soliton states. As earlier, we proceed to solve the PDE by numerical techniques. Considering first the bright solitons, $\rho = 0$, and hence, (2.16) becomes,

$$i\frac{\partial U}{\partial \xi}+\frac{1}{2}\frac{\partial^2 U}{\partial s^2}-\beta\frac{1}{1+|U|^2}U-\alpha\frac{\left(|U|^2\right)}{1+|U|^2}U+\gamma\frac{\left(|U|^2\right)_s}{1+|U|^2}U=0 \tag{2.17}$$

Again, neglecting the effect of diffusion which is plausible if we consider a large value of the photovoltaic and bias field,

$$i\frac{\partial U}{\partial \xi}+\frac{1}{2}\frac{\partial^2 U}{\partial s^2}-\beta\frac{1}{1+|U|^2}U-\alpha\frac{\left(|U|^2\right)}{1+|U|^2}U=0 \tag{2.18}$$

Substituting the bright soliton solution ansatz $U = r^{1/2}y(s)\exp(iv\xi)$, where $r = I(0)/I_d$ and $0 \le y(s) \le 1$ along with the requisite boundary conditions of bright solitons as discussed before in (2.18), we get,

$$\ddot{y}-2vy-2\beta\frac{y}{1+ry^2}+2\alpha\frac{ry^3}{1+ry^2}=0 \tag{2.19}$$

where $\ddot{y} = \frac{d^2y}{ds^2}$.

Integrating (2.19) once and applying the boundary conditions,

$$v=-\frac{\beta+\alpha}{r}\ln(1+r)+\alpha \tag{2.20}$$

$$(\dot{y})^2=2\frac{(\beta+\alpha)}{r}\left[-y^2\ln(1+r)+\ln\left(1+ry^2\right)\right] \tag{2.21}$$

The bright field profile can now found by numerical integration as follows,

$$[2(\beta+\alpha)]^{1/2}s = \pm\int_{y}^{1} \frac{r^{1/2}d\hat{y}}{\left[\ln\left(1+r\hat{y}^2\right)-\hat{y}^2\ln(1+r)\right]^{1/2}} \tag{2.22}$$

In (2.21), we can clearly see that the quantity in square brackets is positive since y(s) is bounded between 0 and 1, and since the LHS is also necessarily positive, we get the condition $\beta+\alpha>0$ for existence of screening photovoltaic bright solitons.

For the dark screening photovoltaic solitons, substituting the appropriate ansatz $U=\rho^{1/2}y(s)\exp(iv\xi)$ along with the dark soliton boundary conditions as stated before, $y(0)=0$, $\dot{y}(0)=0$, $y(s\to\pm\infty)=1$ in (2.16) and neglecting the diffusion effect,

$$\ddot{y}-2vy-2\beta(\rho+1)\frac{y}{1+\rho y^2}-2\alpha\rho\frac{\rho y^3}{1+\rho y^2}=0 \tag{2.23}$$

Using the boundary conditions, one can readily deduce that,

$$v=-\beta \tag{2.24}$$

$$(\dot{y})^2=2(\beta+\alpha)\left[\left(y^2-1\right)-\frac{\rho+1}{\rho}\ln\left(\frac{1+\rho y^2}{1+\rho}\right)\right] \tag{2.25}$$

The soliton field profile can be obtained by numerical integration of (2.25) as follows,

$$[2(\beta+\alpha)]^{1/2}s=\pm\int_{y}^{0}\frac{d\hat{y}}{\left[\left(\hat{y}^2-1\right)-\frac{\rho+1}{\rho}\ln\left(\frac{1+\rho\hat{y}^2}{1+\rho}\right)\right]^{1/2}} \tag{2.26}$$

In (2.25), the quantity in square brackets is positive since *y(s)* is bounded below one, and hence, we can see clearly that $\beta+\alpha<0$ should be satisfied for the RHS to remain positive.

What we can infer from this discussion is that bright or dark solitons can be obtained in photovoltaic photorefractive crystals for suitable conditions. For instance, in Lithium Niobate crystals, where $\alpha<0$, if the external applied bias field is such that $|\beta|<|\alpha|$, then dark solitons can be observed irrespective of the polarity of the external electric field. Again, there are some photovoltaic materials where the photovoltaic constant changes sign under polarization rotation and hence α will be positive or negative depending on the polarization of light. In case $\alpha>0$, the polarity of the external electric field must be reversed to observe dark solitons. Using the values of the Lithium Niobate Crystals as shown in Table 2.1, the normalized intensities of the dark solitons using (2.26) are plotted in Fig. 2.2.

For the grey solitons, we shall employ the grey soliton ansatz, $U(s,\xi)=\rho^{1/2}y(s)\exp\left[i\left(v\xi+\int\frac{Qds}{y^2(s)}\right)\right]$ along with the grey soliton boundary conditions,

Table 2.1 Typical parameters used for lithium niobate crystal in the calculation [3]

Parameter	Value	Parameter	Value
λ_0	0.5 μm	α	−355.13
x_0	40 μm	E_0	(a) 2×10^6 Vm^{-1} (b) -2×10^6 Vm^{-1}
n_e	2.2	β	(a) 177.57, (b) −177.57 (respective to E_0)
r_{33}	30×10^{-12} mV^{-1}	ρ	10
E_p	4×10^6 Vm^{-1}		

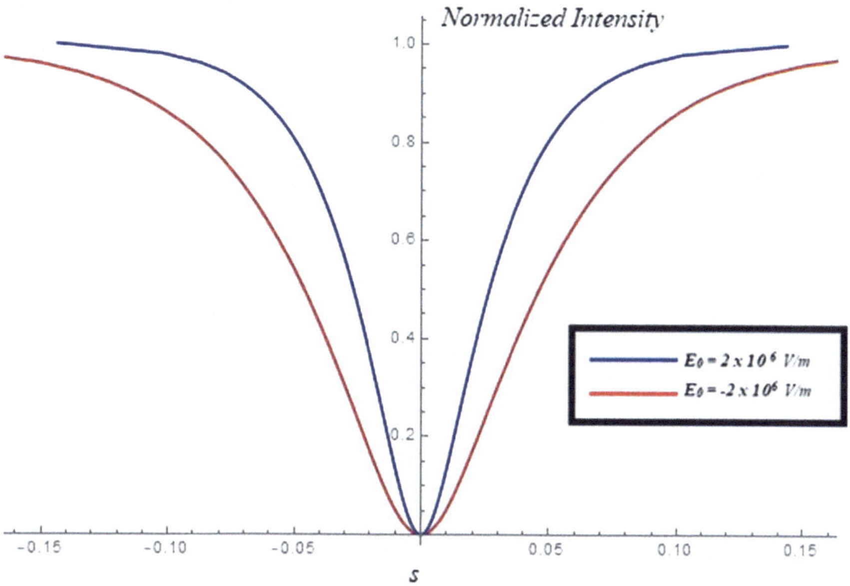

Fig. 2.2 Normalized spatial profiles for the dark screening photovoltaic soliton [3]

$y^2(s=0) = m(0 < m < 1)$, $y(s \to \pm\infty) = 1$, $\dot{y}(0) = 0$. Hence, (2.16) becomes,

$$\ddot{y} - 2\upsilon y - \frac{Q^2}{y^3} - 2\beta(1+\rho)\frac{y}{1+\rho y^2} - 2\alpha\rho\frac{\rho\left(1-y^2\right)y}{1+\rho y^2} = 0 \tag{2.27}$$

or,

$$\ddot{y} - 2(\upsilon - \alpha)y - \frac{Q^2}{y^3} - 2(\alpha+\beta)(1+\rho)\frac{y}{1+\rho y^2} = 0 \tag{2.28}$$

Applying the boundary conditions of the grey solitons at infinity, we have,

$$Q^2 = -2(v + \beta) \tag{2.29}$$

(2.28) yields after an integration,

$$(\dot{y})^2 = 2(v - \alpha)\left(y^2 - 1\right) + 2(v + \beta)\left(\frac{1 - y^2}{y^2}\right) + \frac{2(\alpha + \beta)(1 + \rho)}{\rho}\ln\left(\frac{1 + \rho y^2}{1 + \rho}\right) \tag{2.30}$$

Applying the boundary conditions of grey solitons at zero in (2.30), we get,

$$v = \frac{1}{(m - 1)^2}\left[m(m - 1)\alpha + (m - 1)\beta - \frac{m(\alpha + \beta)(1 + \rho)}{\rho}\ln\left(\frac{1 + \rho m}{1 + \rho}\right)\right] \tag{2.31}$$

From (2.29) and (2.31),

$$(\alpha + \beta)\left[m(m - 1) - \frac{m(1 + \rho)}{\rho} \times \ln\left(\frac{1 + \rho m}{1 + \rho}\right)\right] < 0 \tag{2.32}$$

Expanding $\ln\left(\frac{1+\rho m}{1+\rho}\right) = \ln\left(\frac{1+(m-1)\rho}{1+\rho}\right)$ in series, we can see that the quantity inside the square bracket in (2.32) is positive. That in turn implies the condition for existence of grey solitons in photorefractive photovoltaic media reduces to, $(\alpha + \beta) < 0$. Additionally, the values of (α, β, ρ, m) must be chosen judiciously so as to always have $\dot{y}^2 > 0$ and $Q^2 > 0$. Integrating (2.32) once again gives the spatial profile,

$$[-2(\alpha + \beta)]^{1/2} s = \pm \int_{\sqrt{m}}^{y} (1 - m)\left[\begin{array}{l} \frac{m(1+\rho)}{\rho} \times \ln\left(\frac{1+\rho m}{1+\rho}\right)\left(\tilde{y}^2 + \frac{1}{\tilde{y}^2} - 2\right) + (1 - m)\left(\tilde{y}^2 - 1\right) + \\ m(1 - m)\left(\frac{1}{\tilde{y}^2} - 1\right) - \frac{(m-1)^2(1+\rho)}{\rho}\ln\left(\frac{1+\rho\tilde{y}^2}{1+\rho}\right) \end{array}\right]^{-1/2} d\tilde{y} \tag{2.33}$$

Using (2.33), the normalized spatial profile of the grey soliton is plotted in Fig. 2.3 using typical parameters of the lithium niobate crystal as shown in Table 2.2. Also, unlike the bright and dark screening photovoltaic solitons, the phase is not constant across s. This is evident from the grey soliton ansatz and using (2.29) and (2.31). Figure 2.4 shows the phase profile across the grey soliton.

The screening photovoltaic solitons have a different existential theoretical foundation as compared to screening solitons. There is an inherent interplay of the photovoltaic field with the external bias field. It is interesting to note a few special cases here. If $\alpha = 0$, i.e., we take a non-photovoltaic crystal in (2.16), we retrieve the bright, dark and grey screening solitons formulation. While if we set $\beta = 0$, i.e.,

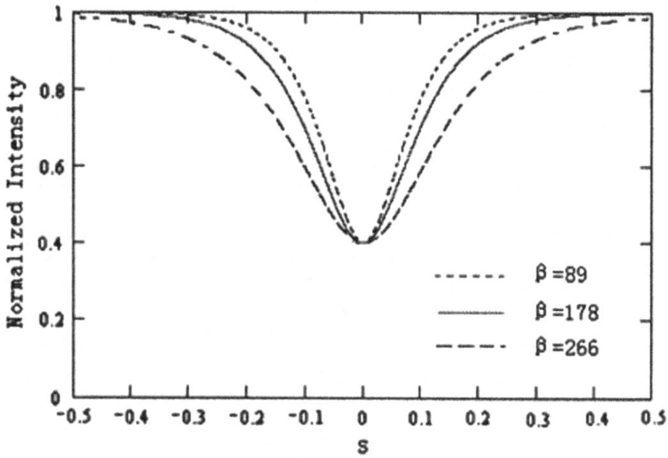

Fig. 2.3 Normalized spatial profiles for the grey screening photovoltaic soliton (reprinted from Optics Communications, 181, Chunfeng Hou, Yan Li, Xiaofu Zhang, Baohong Yuan, Xiudong Sun, Grey screening-photovoltaic spatial soliton in biased photovoltaic photorefractive crystals., 141–144, Copyright 2000, with permission from Elsevier)

Table 2.2 Typical parameters used for lithium niobate crystal in the calculation [3]

Parameter	Value	Parameter	Value
λ_0	0.5 μm	α	−355
x_0	40 μm	E_0	(a) 1×10^6 Vm^{-1} (b) 2×10^6 Vm^{-1} (c) 3×10^6 Vm^{-1}
n_e	2.2	β	(a) 89, (b) 178, (c) 266 (respective to E_0)
r_{33}	30×10^{-12} mV^{-1}	m	0.4
E_p	4×10^6 Vm^{-1}	ρ	5

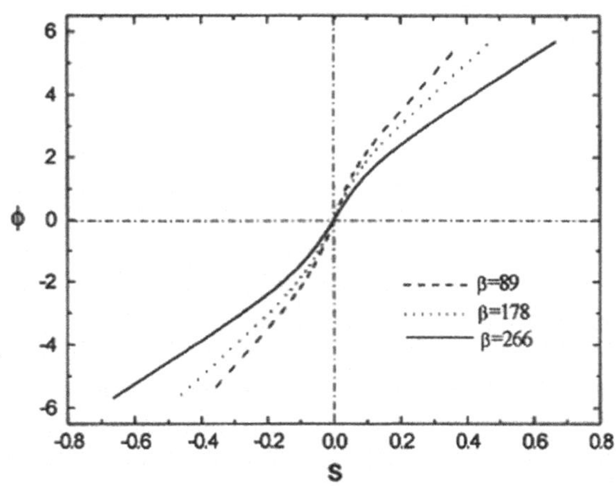

Fig. 2.4 Phase profiles for the grey screening photovoltaic soliton (reprinted from Optics Communications, 181, Chunfeng Hou, Yan Li, Xiaofu Zhang, Baohong Yuan, Xiudong Sun, Grey screening-photovoltaic spatial soliton in biased photovoltaic photorefractive crystals, 141–144, Copyright 2000, with permission from Elsevier)

we switch off the external bias, we obtain the expressions for bright, dark and grey photovoltaic solitons in closed circuit realization ($J \neq 0$ within the crystal) under suitable values of α. For bright photovoltaic solitons, we need $\alpha = r(J+1)/2$ while for dark solitons we need $\alpha = -(J+1)/[2(\rho+1)]$.

2.2.3 *Further Reading*

The reader is referred to [6, 7] where the authors have detailed the photovoltaic solitons. The general formulation presented here branches out to systems of photorefractive crystal which is open circuited or closed circuited and without the external bias field.

2.3 The Pyroelectric Effect

Pyroelectricity is a phenomenon in which a transient voltage is induced in a material when it is heated or cooled. The reason behind this is that when temperature changes momentarily, atoms move around within the crystal lattice and their positions are modified slightly resulting in a change of net polarization. The net polarization change results in a voltage appearing across the crystal. This induced voltage known is known as the pyroelectric voltage and is transient. It will gradually vanish due to leakage current if the temperature remains constant afterward. The leakage current can be present due to a number of causes like movement of electrons through the crystal, movement of ions through the air, or leakage current through a voltmeter attached across the crystal. Contrasting temperature changes induce opposite charges. If heating induces a positive charge on one face, cooling will induce a negative charge at the same face. Quantitatively, pyroelectricity can be said to be the change in net polarization proportional to a change in temperature. The pyroelectric coefficient is defined as,

$$p_i = \frac{\partial P}{\partial T} \tag{2.34}$$

The total pyroelectric coefficient depends upon the primary as well the secondary pyroelectric effect. At constant stress, the piezoelectric contribution from thermal expansion must be added to the pyroelectric coefficients at constant strain to obtain the total pyroelectric coefficient.

All crystal structures can be classified to belong to a set of thirty-two crystal classes, also known as point groups. Twenty-one of these thirty two are non-centrosymmetric. Again, twenty of these twenty one display direct piezoelectricity. In turn, ten of these twenty piezoelectric classes can be expressed to possess a spontaneous polarization, and hence are known as polar classes. Notable is the fact that these also contain a dipole in their unit cell in addition to exhibiting pyroelectricity. Such a material is also ferroelectric if an applied electric field reverses the dipole. In summary, out of the 32 crystal classes, 10 are polar. Since all polar crystals are pyroelectric, so these crystal classes are also known as pyroelectric classes.

2.4 Pyroelectric Solitons (the "Pyroliton")

Pyroelectric solitons, or pyrolitons have been subject of intense research in recent times [8–10]. The transient pyroelectric field alone can cause a stable self trapping as we will see in this section. Combing the transient pyroelectric field with the photovoltaic field or the external electric field can yield interesting effects on the self trapping. There are two conditions which need to be satisfied for the formation of such pyroelectric solitons, firstly that the transient pyroelectric field magnitude is relatively large and secondly, that the the pyroelectric field's relaxation time be greater than the soliton formation time. Assuming the homogeneous heating of a crystal, the pyroelectric field can be expressed as,

$$E_{py} = \frac{1}{\epsilon_0 \epsilon_r} \frac{\partial P}{\partial T} \Delta T \tag{2.35}$$

ΔT is the temperature change. The relaxation time for the pyroelectric field can be given by,

$$\Delta T$$

$$\tau = \frac{\epsilon_0 \epsilon_r}{\sigma_d} \tag{2.36}$$

where σ_d is the dark conductivity of the crystal. Taking typical parameters of the $LiNbO_3$ crystal($\epsilon_r = 28, \sigma_d \sim 10^{-17}\ (\Omega cm)^{-1}$), we find that that the pyroelectric field can remain at significant values for a few weeks. For SBN crystals, the relaxation time is much less as compared to $LiNbO_3$, but this increases considerably if the doping of Ce in SBN crystals > 0.1 wt%. So these are among the most often used crystals which are used for illustration and applications. If we consider a ferroelectric crystal, the distribution of charge present on the crystal faces cancels out the electric field induced due to spontaneous polarization and hence the net field inside a ferroelectric crystal is zero at equilibrium. A temperature change can induces a change in the spontaneous polarization resulting in an electric field E_{py}. This field is not immediately balanced and consequently, a drift current is set up analogous to the effect an external bias has on the crystal. The transient pyroelectric field E_{py} again induces a space charge field which persist to form an index waveguide supporting a soliton. The transient pyroelectric field can replace the external electric field used for screening solitons, with multiple advantages, the main being no need of identifying the *c*-axis since the pyroelectric field automatically manifests along the *c*-axis and secondly, no need of electrodes on the crystal among others [11, 12].

2.4.1 Theoretical Formulation

Consider a light beam propagating along the *z*-axis and assume the diffraction only in x-direction. The soliton beam is polarized along the positive *x*-direction. The crystal is kept such that its *c*-axis coincides with the positive *x* axis. A temperature controlled (via a Peltier cell) metal plate is kept in contact with the crystal. In addition, a thermally insulating cover is kept on top of the crystal for minimizing undesirable external effects on the temperature. The incident beam can be stated as a slowly varying envelope $\boldsymbol{E} = \hat{x}A(x, z)\exp(ikz)$ where $k = k_0 n_e$, n_e is the unperturbed refractive index, n_e' is the refractive index along the *c*-axis and λ_0 is the free space wavelength. Under these assumptions, the paraxial diffraction equation for the dynamical evolution becomes,

$$i\frac{\partial\phi}{\partial z} + \frac{1}{2k}\frac{\partial^2\phi}{\partial x^2} - \frac{k_0 n_e^3 r_{eff} E_{pysc}}{2}\phi = 0 \tag{2.37}$$

where r_{eff} is the electro-optic coefficient, E_{pysc} is the space charge field induced solely by the pyroelectric effect [8]. It is essential to now obtain an expression for the induced space charge field E_{pysc} due to the pyroelectric effect. To this end, Ohm's law in differential form can be stated as,

$$\vec{j} = \sigma\vec{E} \tag{2.38}$$

The continuity equation is,

$$\frac{\partial\rho}{\partial t} + \nabla.\vec{j} = 0 \tag{2.39}$$

While the Gauss Law states,

$$\nabla\cdot\vec{D} = \rho \tag{2.40}$$

where $\vec{j}$ is the total current, $\sigma = \kappa I + \sigma_d$ is the total conductivity, E is the total electric field. P is the space charge field density, κ is the specific photoconductivity and D is the electric displacement. The light intensity is $I = (n_e/2\eta_0)|\phi|^2$ and is a function of x being expressed as, $I(x) = I_0\exp\left[-2(x/x_1)^2\right]$. x_1 is the characteristic beam radius and I_0 is the maximum intensity at the beam center. Then, the total conductivity will be stated as, $\sigma = \sigma_0\left[\exp\left[-2(x/x_1)^2\right] + \eta\right]$ with $\sigma_0 = \kappa I_0$ and $\eta = I_d/I_0$ and I_{d} is the dark irradiance. Since we consider an SBN crystal in open circuit, and considering the illuminated region to be narrow compared to the thickness of the crystal, total current j can be expressed as,

$$j = j_d = \sigma_d \frac{V}{H} = \sigma_d E_{py} \tag{2.41}$$

where j_d is the divergence less current which satisfies the boundary conditions. Solving (2.38)-(2.40), we get,

$$\nabla \cdot \left[\epsilon_0 \epsilon_r \frac{\partial \vec{E}}{\partial t} + \sigma \vec{E} \right] = 0 \tag{2.42}$$

Considering the boundary conditions and assuming negligible effect of diffusion and photovoltaic effects,

$$\epsilon_0 \epsilon_r \frac{\partial \vec{E}}{\partial t} + \sigma \vec{E} = \vec{j}_d \tag{2.43}$$

Solving the partial differential Eq. (2.43),

$$E(\bar{t}, x_1) = \frac{V}{H} \left\{ \begin{array}{l} \frac{\eta}{\exp(-2x^2/x_1^2)+\eta} + \\ \frac{\exp(-2x^2/x_1^2)}{\exp(-2x^2/x_1^2)+\eta} \exp\left[-\bar{t}\left(\exp(-2x^2/x_1^2) + \eta\right)\right] \end{array} \right\} \tag{2.44}$$

where $\bar{t} = t/t_d$, $\tau = \epsilon_0 \epsilon_r / \sigma_0$ is known as the characteristic Maxwell time. Now, two components constitute the total electric field, $E = E_{py} + E_{pysc}$. E_{py} is the homogeneous pyroelectric field induced by the homogeneous heating. This causes a homogeneous refractive index change. E_{pysc} is the inhomogeneous space charge field which causes an inhomogeneous refractive index change. The origin of self trapping lies in this refractive index waveguide. Hence,

$$E_{pysc} = E - E_{py} = E_{py} \frac{\exp(-2x^2/x_1^2)}{\exp(-2x^2/x_1^2) + \eta} \left\{ \exp\left[-\bar{t}\left(\exp(-2x^2/x_1^2) + \eta\right)\right] - 1 \right\} \tag{2.45}$$

For $\bar{t} = 0$, it is plain that $E_{\mathrm{pysc}} = 0$ which implies that pyroelectric field has not been screened yet in the illuminated region. At steady state, we know $\bar{t} \gg 1$ and hence the terms in the curly brackets in (2.45) tends to -1,

$$E_{pysc} = -E_{py} \frac{I}{I + I_d} \tag{2.46}$$

The value of the pyroelectric space charge field is dependent upon E_{py} and hence the change in temperature ΔT.The expression is also similar to the space charge field in open circuit photovoltaics.

Substituting (2.46) into (2.37), we have,

$$i\frac{\partial U}{\partial \xi}+\frac{1}{2}\frac{\partial^2 U}{\partial s^2}+\alpha\frac{|U|^2}{1+|U|^2}U=0 \tag{2.47}$$

where we have used the usual dimensionless coordinates, $\xi = z/\left(kx_0^2\right)$, $s = x/x_0$, $\phi = (2\eta_0 I_d/n_e)^{1/2}U$ with x_0 to be an arbitrary spatial width and the intensity scaled with the dark irradiance I_d, $\rho = I_\infty/I_d$, $\alpha = (k_0 x_0)^2\left(n_e^4 r_{eff}/2\right)E_{py}$.

2.4.2 *Bright, Dark and Grey Solitons*

For the solution of the bright solitons, it is now straightforward with the above theoretical foundation. Using the bright soliton ansatz, $U = r^{1/2}y(s)\exp(i v\xi)$ with the bright soliton boundary conditions, $y(0) = 1$, $\dot{y}(0) = 0$, $y(s \to \pm\infty) = 0$. $y(s)$ a bounded function such that $0 \le y(s) \le 1$ and $r = I(0)/I_d$. Substituting the bright soliton ansatz in (2.47), we get,

$$\frac{d^2 y}{ds^2}=2vy-2\alpha\frac{ry^3}{1+ry^2} \tag{2.48}$$

Integrating (2.48), we can obtain the soliton field profile,

$$s=\pm\int_y^1\left\{\frac{2\alpha}{r}\left[\ln\left(1+r\tilde{y}^2\right)-\tilde{y}^2\ln(1+r)\right]\right\}^{-1/2}d\tilde{y} \tag{2.49}$$

For the dark soliton solution, using the equivalent dark soliton ansatz $U = \rho^{1/2}y(s)\exp(i\mu\xi)$ along with the boundary conditions, we substitute in (2.47) and integrate to obtain for the spatial profile,

$$s=\pm\int_y^1\left\{-2\alpha\left[\frac{y^2-1}{1+\rho}-\frac{1}{\rho}\ln\left(\frac{1+\rho\tilde{y}^2}{1+\rho}\right)\right]\right\}^{-1/2}d\tilde{y} \tag{2.50}$$

Similarly, for the grey soliton, we shall substitute, $U(s,\xi) = \rho^{1/2}y(s)\exp\left[i\left(\mu\xi+\int\frac{Qds}{y^2(s)}\right)\right]$ into (2.47) and integrate once to yield,

Table 2.3 Typical factors of lithium niobate crystal used in our calculation [8]

Parameter	Value	Parameter	Value
λ_0	532 nm	ε_0	8.85×10^{-12} F/m
x_0	20 μm	ε_r	3400
n_e	2.35	$\frac{\partial P}{\partial T}$	-3×10^{-4} Cm^{-2} K^{-1}
r_{eff}	237×10^{-12} mV^{-1}	r,ρ	10

$$\left(\frac{dy}{ds}\right)^2 = 2\mu\left(y^2 - 1\right) - \left(\frac{1}{y^2} - 1\right)Q^2 - \frac{2\alpha}{\rho}\left[\rho\left(y^2 - 1\right) - \ln\left(\frac{1 + \rho y^2}{1 + \rho}\right)\right] \tag{2.51}$$

The boundary conditions are, $y^2(s = 0) = m(0 < m < 1)$, $\dot{y}(0) = 0$, $y(s \to \pm\infty) = 1$.

Using the boundary conditions, we can easily find,

$$Q^2 = 2\alpha\frac{\rho}{1 + \rho} - 2\mu \tag{2.52}$$

$$\mu = \frac{1}{2(m-1)^2}\left\{(1 - m)\left(\frac{2\alpha\rho}{1 + \rho}\right) + \frac{2m\alpha}{\rho}\left[\rho(m - 1) - \ln\left(\frac{1 + \rho m}{1 + \rho}\right)\right]\right\} \tag{2.53}$$

The normalized intensity profile of the grey soliton can be obtained by numerically integrating (2.51) along with (2.52) and (2.53). Table 2.3 shows the typical parameters used for calculation. Using this and (2.48)–(2.52), the bright, dark and grey soliton intensity profiles have been plotted in Figs. 2.5,2.6,2.7 (Table 2.4).

We can see clearly in (2.49) that the quantity within brackets is positive only if $\alpha > 0$. So we need to take the change of temperature as positive, i.e., $\Delta T > 0$ and we need a heating of the crystal. Similarly, for the dark solitons, the term inside brackets in

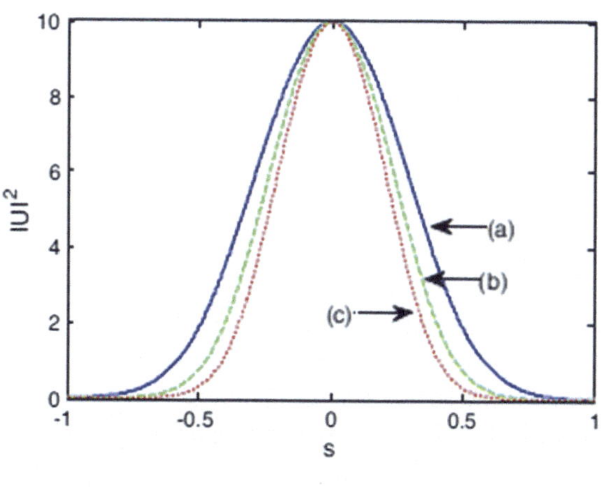

Fig. 2.5 Normalized spatial profile of the bright soliton for **a** $\Delta T = 10$ °C **b** $\Delta T = 20$ °C **c** $\Delta T = 30$ °C (reprinted from Optik, 126, Yanli Su, Qichang Jiang, Xuanmang Ji, Photorefractive spatial solitons supported by pyroelectric effects in strontium barium niobate crystals, 1621–1624, Copyright 2015, with permission from Elsevier)

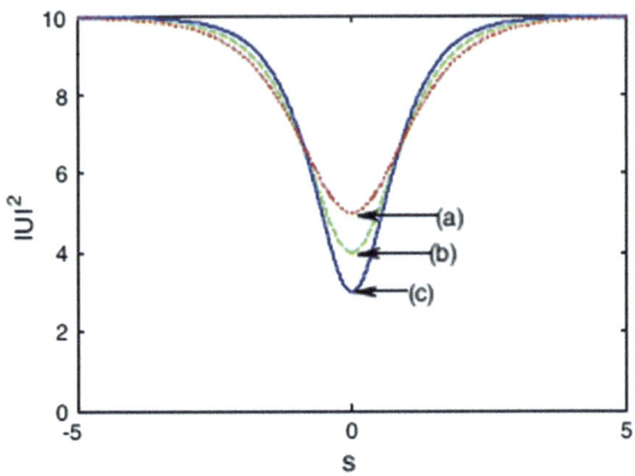

Fig. 2.6 Normalized spatial profile of the grey soliton for **a** $\Delta T = -10$ °C, $m = 0.5$ **b** $\Delta T = -10$ °C, $m = 0.4$ **c** $\Delta T = -10$ °C, $m = 0.3$ (reprinted from Optik, 126, Yanli Su, Qichang Jiang, Xuanmang Ji, Photorefractive spatial solitons supported by pyroelectric effects in strontium barium niobate crystals, 1621–1624, Copyright 2015, with permission from Elsevier)

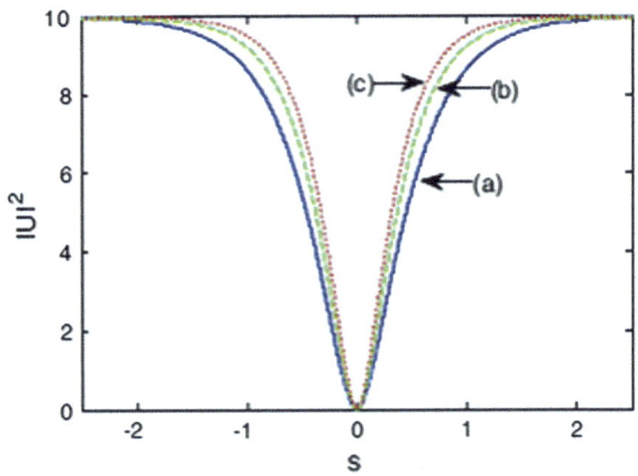

Fig. 2.7 Normalized spatial profile of the dark soliton for **a** $\Delta T = -10$ °C **b** $\Delta T = -20$ °C **c** $\Delta T = -30$ °C. (reprinted from Optik, 126, Yanli Su, Qichang Jiang, Xuanmang Ji, Photorefractive spatial solitons supported by pyroelectric effects in strontium barium niobate crystals, 1621–1624, Copyright 2015, with permission from Elsevier)

Table 2.4 Values of α for diverse values of the temperature change using Table 2.3

ΔT (°C)	α
10	20.1
20	30.2
30	40.2
−10	−20.1
−20	−30.2
−30	−40.2

(2.50) is positive only if $\alpha < 0$. Hence, we need a cooling of the crystal with $\Delta T < 0$ for observing dark solitons. The same concept carries forward for grey solitons also where again we need a cooling of the crystal for their observation. The important thing to note here is that the nonlinearity is controlled by the term α which is in turn dependent upon the temperature change, pyroelectric coefficient and electro-optic coefficient. Varying these values for different types of crystals can result in different characteristics and spatial profiles of the pyrolectric solitons. (Fig. 2.6)

2.5 Photovoltaic Effect and Pyroelectric Solitons

Photorefractive solitons observed in steady state can be said to be broadly of three types, i.e., screening solitons, photovoltaic solitons and screening–photovoltaic solitons [3, 4, 7].

An external electric field leads to a screening of the induced space charge field and hence the name "screening solitons". The photorefractive effect is basically a refractive index change by the electro-optic effect. The electro-optic effect comes into play because of the induced space charge field due to the drift and diffusion of photogenerated charge carriers. In case of photovoltaic solitons, the space charge field is modulated by the bulk photovoltaic field while screening photovoltaic solitons form in photorefractive photovoltaic crystals due to the combination of both the external field and bulk photovoltaic field. As we have mentioned before, replacing the external electric field with the pyroelectric field has many advantages. It is logical to now think of the combination of the external bias field, photovoltaic field and pyroelectric field and how they can have an interplay while inducing a space–charge field and in turn self trapping a light beam. Such type of solitons are screening photovoltaic pyroelectric solitons. Also, the transient pyroelectric field can be induced by externally controlled temperature changes or by absorption of the incident beam's energy. The former case has already been seen in the previous section. In the following, we now discuss the effect of pyroelectricity due to the absorption of energy of the beam itself.

2.5.1 *Theoretical Model*

We consider the usual setup for the soliton beam as defined before. In addition, the crystal is covered with a thermally insulating cover so as to stabilize the temperature and avoid any temperature gradient. The slowly varying envelope for the incident beam reads as, $\boldsymbol{E} = \hat{x}A(x,z)exp(ikz)$ where $k = k_0 n_e$, n_e is the unchanged refractive index, n_e' is the perturbed refractive index along the direction of the extraordinary or c-axis and λ_0 is the wavelength in free space. Beginning with the usual paraxial diffraction equation,

$$\left(i\frac{\partial}{\partial z} + \frac{1}{2k}\frac{\partial^2}{\partial x^2} + \frac{k}{n_e}\Delta n\right)A(x,z) = 0 \tag{2.54}$$

$$\Delta n = -\frac{1}{2}n_{\mathrm{e}}^3 r_{\mathrm{eff}} E_{\mathrm{sc}} \tag{2.55}$$

Here, E_{sc} is the space charge field resulting as a consequence of both, the photovoltaic drift and the pyroelectric field. Again there are three constituent components of the total space charge field, which are the space charge field due to the external bias, the photovoltaic contribution and the pyroelectric space charge field,

$$E_{sc} = E_1 + E_2 + E_3 \tag{2.56}$$

The space charge field ($E_1 + E_2$) results in biased photorefractive crystals supporting screening photovoltaic solitons. We have already studied it in detail in the previous section,

$$E_1 + E_2 = E_0\frac{I_\infty + I_d}{I + I_d} + E_p\frac{I_\infty - I}{I + I_d} \tag{2.57}$$

The value of E_p is reliant on the state of polarization of the beam and one can infer the sign from the photovoltaic constant. E_{pysc} is the space–charge field forming as a consequence of the transient pyroelectric field E_{py} which results in turn from a change in temperature. A pulse of light can transfer energy to the material and hence induce the pyroelectric effect comparable to that brought about by a change in temperature [8, 10, 13–16]. For a short pulse of light, the pyroelectric space charge field E_{pysc} can be expressed as, [8, 10, 13–16],

$$E_3 = E_{pysc} = -\frac{1}{\varepsilon_r\varepsilon_0}\frac{\partial P_s}{\partial T}\frac{t_p}{2}\frac{\sigma_{ph}}{\varepsilon_0\varepsilon_r}\Delta T(t) = -E_{py}\frac{t_p}{2}\frac{\sigma_{ph}}{\varepsilon_0\varepsilon_r} \tag{2.58}$$

Now, the value of E_{pysc} is implicitly a function of the intensity of the beam. For our calculation, we need an explicit dependence and hence we approximate the space charge field as [8, 10],

$$E_{pysc} = -E_{py}\frac{t_p}{2}\frac{\sigma_{ph}}{\varepsilon_0\varepsilon_r} \approx -E_{py}\frac{\vartheta I}{I_d} \tag{2.59}$$

where, t_p is the pulse duration, σ_{ph} is the photoconductivity, ϑ is a material parameter dependent on the crystal. Since the photoconductivity σ_{ph} is proportional to the intensity I., this approximation is reasonable. From the values of the parameters E_{py}, t_p,ε_0, ε_r [8, 10, 13–16], we can say that $\frac{\lambda I}{I_d} < 1$. Again, this can be verified independently as the pyroelectric space charge field can reach a substantial fraction of the pyroelectric field a under continuous wave laser beam [15]. The following potential condition in steady state can be used to find the value of E_0,

$$-\int_{-l/2}^{l/2} E_{sc}dx = \varepsilon \tag{2.60}$$

where the transverse thickness of the crystal is represented by l, and the external bias is denoted by ε. Substituting (2.57)–(2.60) in (2.56), we get,

$$E_{\mathrm{sc}} = -(\varepsilon\eta + E_{\mathrm{p}}\sigma\eta - E_{\mathrm{py}}\gamma\lambda\eta)\frac{I_\infty + I_d}{I + I_d} + E_{\mathrm{p}}\frac{I_\infty - I}{I + I_d} - E_{\mathrm{py}}\frac{\lambda I}{I_d} \tag{2.61}$$

where $\eta = \frac{1}{\int_{-l/2}^{l/2} \frac{I_\infty + I_d}{I + I_d} dx}$, $\sigma = \int_{-l/2}^{l/2} \frac{I_\infty - I}{I + I_d} dx$ and $\gamma = \int_{-l/2}^{l/2} \frac{I}{I_d} dx$.

Using (2.61) in the paraxial diffraction equation, we obtain the dynamical evolution equation as follows,

$$iU_\xi + \frac{1}{2}U_{ss} + \beta(|U|^2)U - \frac{\alpha(\rho - (|U|^2)}{(1 + |U|^2)}U - \delta\frac{(1+\rho)}{(1 + |U|^2)}U = 0 \tag{2.62}$$

where we have used the previously defined dimensionless coordinates and, $\beta = E_{\mathrm{py}}$, $\tau = (k_0 x_0)^2 n_e^4 r_{\mathrm{eff}}/2$, $\alpha = \tau E_p$, $\delta = -(\varepsilon\eta + E_p\sigma\eta - E_{\mathrm{py}}\gamma\lambda\eta)\tau$.

Other symbols have their meaning as defined before.

2.5.2 *Spatial Soliton States*

Using the bright soliton ansatz, $U = r^{\frac{1}{2}} f(s) \exp(i\mu\xi)$ and substituting in (2.62),

$$\ddot{f} = 2\mu f - 2\beta(rf^2)f - 2\alpha\frac{(rf^2)f}{1 + rf^2} + 2\delta\frac{f}{1 + rf^2} \tag{2.63}$$

where $\ddot{f} = \frac{d^2 f}{ds^2}$.

Integrating (2.63), and by use of the boundary conditions for bright solitons,

$$\dot{f}^2 = 2\mu f^2 - \beta r f^4 + \frac{(2(\delta + \alpha)\log(1 + rf^2) - 2\alpha r f^2)}{r} + c \tag{2.64}$$

with

$$c = 0 \tag{2.65}$$

$$\mu = \frac{\beta r}{2} - \left(\frac{\delta}{r} + \frac{\alpha}{r}\right)\log(1 + r) + \alpha \tag{2.66}$$

Integrating (2.14) once again, we get the envelope,

$$s = \pm \int_{f}^{1} \frac{r d\tilde{f}}{2\mu r \tilde{f}^2 - \beta r^2 \tilde{f}^4 + 2(\delta + \alpha)\log\left(1 + r\tilde{f}^2\right) - 2\alpha r \tilde{f}^2} \tag{67a}$$

As $\dot{f}$ has to be real and bounded like $0 \leq f(s) \leq 1$, the sufficient condition to keep RHS positive can be inferred from (2.64),

$$-\frac{\beta r}{2} + \mu - \alpha \geq -\frac{\delta + \alpha}{r}\log(1 + r) \tag{67b}$$

For bright spatial solitons, we need positive refractive index perturbation. If we consider a lithium niobate crystal for illustration, the refractive index perturbation is negative due to there being a self-defocussing due to its negative photovoltaic coefficient. In open circuit crystals, a bright soliton can still be self trapped in the event that the self focusing induced due to the pyroelectric effect more than offsets the self-defocussing due to the photovoltaic effect [9].

If an external bias is applied, this condition is liable to be modified slightly. An external electric field applied along the *c*-axis results in self focussing while leading to self defocussing if applied in the opposite direction [4]. Hence, the photovoltaic self defocussing can be boosted or reduced by controlling the direction of the voltage bias. This, in turn has the tendency to change the degree of self focusing induced by the pyroelectric effect. In summary, these changes to the nonlinearity will result in an alteration of the FWHM of the soliton.

Now, with regards to the wavelength of incident light, we need to consider a wavelength for which absorption of energy is relatively large so that the heating of the crystal can take place [10]. The light absorption in $LiNbO_3$ is significant for the blue-violet light ~405 nm [11] and hence will be used in our simulation. Now, the photoconductivity increases substantially when considering light in the blue violet region as compared to the red light and hence the photovoltaic field also decreases substantially as it is inversely proportional to the photoconductivity [17–19]. Also, the photovoltaic field has a sub-linear dependence on the incident light intensity in case of undoped $LiNbO_3$[20]. At the chosen wavelength and intensity we consider, we should use a lesser value of E_p than that considered before for $LiNbO_3$ (in Sect. 2.4, as in [3, 6, 7]). So, a judicious conjecture would be $E_p = -2 \times 10^5$ V/m [10]. The transient pyroelectric field is approximately taken to be ~40 kV/cm, which correlates to a temperature change of the order of around ~10 K in the crystal [12]. To summarize [10], $\lambda_0 = 405$ nm, $x_0 = 20$ μm., $E_p = -2 \times 10^5$ V/m ~ -2 kV/cm, $E_{\mathrm{py}} = 4.0 \times 10^6$ V/m, $r_{\mathrm{eff}} = r_{33} \sim 35 \times 10^{-12}$ m V^{-1}, $n_e = 2.2$, $\lambda = 0.5$, $r = 10$.

With the above parameters, we obtain the value of $\alpha = -6.846$ and $\beta = 67.275$. The soliton profiles for an applied voltage $\varepsilon = \pm\ 4000$ V are shown in Fig. 2.8. The thickness of the crystal $l = 10$ mm. Again, if we consider (hypothetically) the photovoltaic field constant $E_{\mathrm{p}} = +\ 2$ kV/cm and rest of the parameters same as before, we find that the photovoltaic effect will work to support the pyroelectric effect augmenting the self trapping. This case is shown in Fig. 2.9. Hence, the interaction

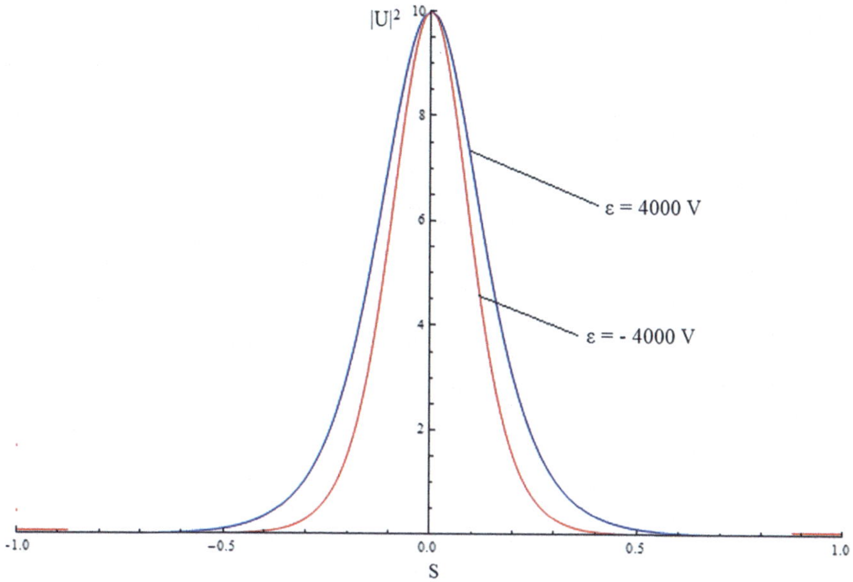

Fig. 2.8 Spatial intensity profile of the solitons when $\alpha = -6.846$, $\beta = 67.275$, $r = 10$ (reprinted from Physics Letters A, 381, Aavishkar Katti, R.A. Yadav, Spatial solitons in biased photovoltaic photo refractive materials with the pyroelectric effect, 166–170, Copyright 2017, with permission from Elsevier)

between the photovoltaic effect, external bias and the pyroelectric field can be clearly seen [10].

For the dark soliton, we take the field profile, as usual, $U = \rho^{\frac{1}{2}} g(s) \exp(i\nu\xi)$ where g(s) is a bounded function and along with the boundary conditions of dark solitons as specified before. Substituting the dark soliton ansatz in in (2.62), we get,

$$\ddot{g} = 2\nu g - 2\beta(\rho g^2)g + 2\alpha\frac{(\rho - \rho g^2)g}{1 + \rho g^2} + 2\delta\frac{(\rho + 1)g}{1 + \rho g^2} \tag{2.68}$$

where $\ddot{g} = \frac{d^2 g}{ds^2}$.

By using the boundary conditions at infinity in (2.68),

$$\nu = \beta\rho - \delta \tag{2.70}$$

Integrating (2.68) once,

$$\dot{g}^2 = 2\nu g^2 - \beta\rho g^4 + \frac{2\delta(1 + \rho)\log(1 + \rho g^2)}{\rho} - 2\alpha g^2$$

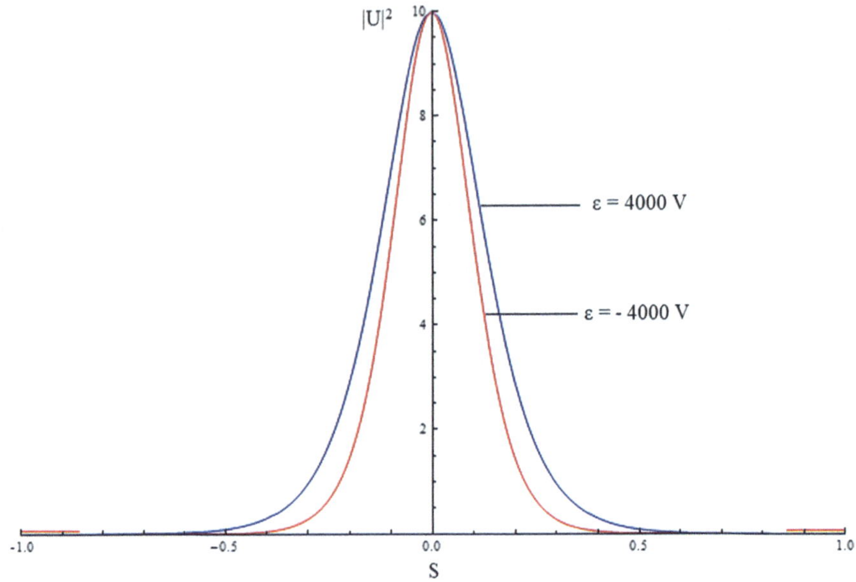

Fig. 2.9 Spatial intensity profile of the solitons when $\alpha = 6.846$, $\beta = 67.275$, $r = 10$ (reprinted from Physics Letters A, 381, Aavishkar Katti, R.A. Yadav, Spatial solitons in biased photovoltaic photo refractive materials with the pyroelectric effect, 166–170, Copyright 2017, with permission from Elsevier)

$$+\frac{2\alpha(1+\rho)\log\left(1+\rho g^2\right)}{\rho}+2c \tag{2.71}$$

Using the boundary conditions in (2.71) at infinity, we can obtain c,

$$c=-\nu+\frac{\beta\rho}{2}+\alpha-\frac{(\alpha+\delta)(1+\rho)\log(1+\rho)}{\rho} \tag{2.72}$$

Integrating (2.71) using the value of c by (2.72), the soliton envelope can be derived as,

$$s=\pm\int\limits_0^g \frac{\rho d\tilde{g}}{2\nu\rho\tilde{g}^2-\beta\rho^2\tilde{g}^4+2(\delta+\alpha)(1+\rho)\log\left(1+\rho\tilde{g}^2\right)-2\alpha\rho\tilde{g}^2+2\rho c} \tag{2.73}$$

Analogous to the case of bright solitons, the refractive index change is expected to be negative for self-defocussing to happen and dark solitons to form. The photovoltaic induced self defocussing should counteract the self-focussing induced by the the external bias field and the pyroelectric effect for a dark soliton to form. So, it is apt to consider the external bias values as $\varepsilon = 4000$ V and $\varepsilon = 40{,}000$ V. If the voltage bias is

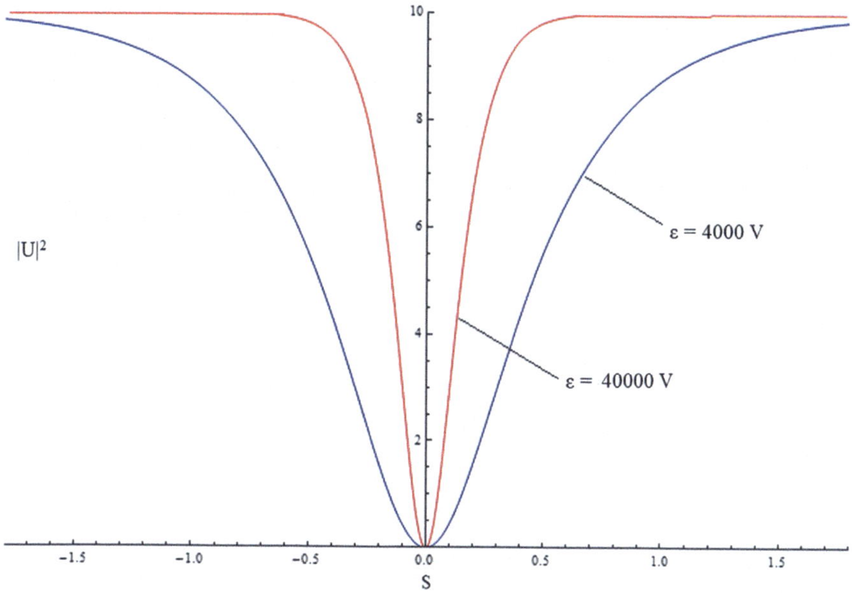

Fig. 2.10 The normalized intensities of the solitons when $\alpha = -6.846$, $\beta = 6.7275$, $\rho = 10$ (reprinted from Physics Letters A, 381, Aavishkar Katti, R.A. Yadav, Spatial solitons in biased photovoltaic photorefractivematerials with the pyroelectric effect, 166–170, Copyright 2017, with permission from Elsevier)

negative, the electric field is parallel to the c-axis. So the overwhelming self focussing effect due to the induced pyroelectric space charge field prevents self defocussing. Hence, we consider [10], the following parameters, $\lambda_0 = 405$ nm, $x_0 = 20$ µm., $E_p = -2.0 \times 10^5$ V/m, $E_{py} = 4.0 \times 10^5$ V/m, $r_{eff} = 35 \times 10^{-12}$ m V^{-1}, $n_e = 2.2$, $\lambda = 0.5$, $\rho = 10$. Also, we have, $\alpha = -6.846$, $\beta = 6.7275$. The soliton profiles are shown in Fig. 2.10.

2.6 Concluding Remarks and Further Reading

We have studied a comprehensive theory for optical spatial solitons in photovoltaic and pyroelectric photorefractive crystals. The band transport model is used to obtain the space charge field due to photovoltaic effects and the dynamical evolution equation for light beams propagating in biased photovoltaic photorefractive crystals. The general formulation presented here branches out to systems of photorefractive crystal which is open circuited or closed circuited and without the external bias field. The pyroelectric space charge field and the dynamical evolution equation is also obtained using simple considerations of charge transport and continuity equations. Screening photovoltaic pyroelectric solitons are studied to investigate the interplay between the pyroelectric field, external bias and photovoltaic field.

Buse [13–16] give an understanding of the different characteristics of the pyroelectric effect in photorefractive crystals. Since we focus on the theory behind the self trapping due to different configurations of pyroelectric photorefractive crystals, the author is referred to [8, 9, 11, 21] for further details of pyroliton in the experimental context. [10] has detailed exposition of screening photovoltaic pyroelectric solitons.

References

1. P. Gunter, J.-P. Huignard (eds.), *Photorefractive Materials and Their Applications 1* (Springer Science+Business Media Inc., New York, 2006)
2. D.D. Nolte, *Photorefractive Effects and Materials*, vol. 27 (1995)
3. J.S. Liu, K.Q. Lu, Screening-photovoltaic spatial solitons in biased photovoltaic-photorefractive crystals and their self-deflection. J. Opt. Soc. Am. B-Opt. Phys. **16**(4), 550–555 (1999)
4. D.N. Christodoulides, M.I. Carvalho, Bright, dark, and gray spatial soliton states in photorefractive media. J. Opt. Soc. Am. B **12**(9), 1628 (1995). https://doi.org/10.1364/JOSAB.12.001628
5. N.V. Kukhtarev, V.B. Markov, S.G. Odulov, M.S. Soskin, V.L. Vinetskii, Holographic storage in electrooptic crystals: I. Steady state. Ferroelectrics **22**(1), 949–960 (1978). https://doi.org/10.1080/00150197908239450
6. G.C. Valley, M. Segev, B. Crosignani, A. Yariv, M.M. Fejer, M.C. Bashaw, Dark and bright photovoltaic spatial solitons. Phys. Rev. A **50**(6), R4457 (1994). https://doi.org/10.1103/PhysRevA.50.R4457
7. M. Segev, G.C. Valley, M.C. Bashaw, M. Taya, M.M. Fejer, Photovoltaic spatial solitons. J. Opt. Soc. Am. B **14**(7), 1772 (1997). https://doi.org/10.1364/JOSAB.14.001772
8. Y. Su, Q. Jiang, X. Ji, Photorefractive spatial solitons supported by pyroelectric effects in strontium barium niobate crystals. Optik **126**(18), 1621–1624 (2015). https://doi.org/10.1016/j.ijleo.2015.04.053
9. J. Safioui, F. Devaux, M. Chauvet, Pyroliton: pyroelectric spatial soliton. Opt. Express **17**(24), 22209–22216 (2009). https://doi.org/10.1364/OE.17.022209
10. A. Katti, R.A. Yadav, Spatial solitons in biased photovoltaic photorefractive materials with the pyroelectric effect. Phys. Lett. Sect. A: Gen. At. Solid State Phys. **381**(3), 166–170 (2017). https://doi.org/10.1016/j.physleta.2016.10.054
11. S.T. Popescu, A. Petris, V.I. Vlad, Fast writing of soliton waveguides in lithium niobate using low-intensity blue light. Appl. Phys. B: Lasers Opt. **108**(4), 799–805 (2012). https://doi.org/10.1007/s00340-012-5202-7
12. S.T. Popescu, A. Petris, V.I. Vlad, Recording of self-induced waveguides in lithium niobate at 405 nm wavelength by photorefractive–pyroelectric effect. J. Appl. Phys. **113**(21), 213110 (2013). https://doi.org/10.1063/1.4808321
13. K. Buse, Light-induced charge transport processes in photorefractive crystals I: models and experimental methods. Appl. Phys. B: Lasers Opt. **64**(3), 273–291 (1997). https://doi.org/10.1007/s003400050175
14. K. Buse, K.H. Ringhofer, Pyroelectric drive for light-induced charge transport in the photorefractive process. Appl. Phys. A Solids Surf. 57(2), 161–165 (1993). https://doi.org/10.1007/BF00331438
15. K. Buse, R. Pankrath, E. Krätzig, Pyroelectrically induced photorefractive effect in Sr(0.61)Ba(0.39)Nb(2)O(6):Ce. Opt. Lett. **19**(4), 260–262 (1994). https://doi.org/10.1364/OL.19.000260

16. N. Korneev, D. Mayorga, S. Stepanov, A. Gerwens, K. Buse, E. Krätzig, Enhancement of the photorefractive effect by homogeneous pyroelectric fields. Appl. Phys. B **66**, 393 (1998)
17. J.R. Schwesyg, M.C.C. Kajiyama, M. Falk, D.H. Jundt, K. Buse, M.M. Fejer, Light absorption in undoped congruent and magnesium-doped lithium niobate crystals in the visible wavelength range. Appl. Phys. B: Lasers Opt. **100**(1), 109–115 (2010). https://doi.org/10.1007/s00340-010-4063-1
18. M. Alonzo et al., Self-confined beams in erbium-doped lithium niobate. J. Opt. A: Pure Appl. Opt. **12**(1), 015206 (2010). https://doi.org/10.1088/2040-8978/12/1/015206
19. F. Lüdtke, N. Waasem, K. Buse, B. Sturman, Light-induced charge-transport in undoped LiNbO 3 crystals. Appl. Phys. B: Lasers Opt. **105**(1), 33–50 (2011). https://doi.org/10.1007/s00340-011-4615-z
20. R. Jungen, G. Angelow, F. Laeri, and C. Grabmaier, Efficient ultraviolet photorefraction in LiNbO3. Appl. Phys. A Solids Surf. **55**(1), 101–103 (1992). https://doi.org/10.1007/BF00324609
21. S.T. Popescu, A. Petris, V.I. Vlad, Recording of self-induced waveguides in lithium niobate at 405 nm wavelength by photorefractive-pyroelectric effect. J. Appl. Phys. **113**(21), 2013. https://doi.org/10.1063/1.4808321

Chapter 3
Stability and Dynamical Evolution

3.1 Introduction

Modulation instability is a distinctive feature associated with most nonlinear dispersive systems. It is a very common occurrence in nonlinear optics and fluid dynamics [1–9]. Modulational instability, also known as sideband instability can be said to be a phenomenon where certain perturbations of a periodic waveform get strengthened consequently leading to the generation of spectral-sidebands and the ensuing breakup of the waveform into small filaments. The growth and evolution of the periodic disturbances or noise on a continuous wave background has to be exponential this makes modulation instability a form of amplification. It was first observed by T. Brooke Benjamin and Jim E. Feir, in 1967 in case of periodic surface gravity waves on deep water known [1, 7]. Hence, it is also known as Benjamin – Feir instability.

The instability is reliant on the frequency of the perturbation. It is possible that at certain frequencies, the perturbation attenuates having a negligible effect on the beam propagation. While at some other frequencies, the perturbation grows exponentially resulting in breakup of the light beam. The complete gain spectrum can be obtained analytically by the standard theoretical paraxial Helmholtz equation solved for perturbed optical beam. Random perturbations contain a wide range of frequencies and so the underlying gain spectrum has spectral sidebands.

Modulation Instability is said to herald the process of soliton formation. It shares the same parameter space as that of spatial soliton formation [2]. The primary cause is due to the local and global effects of the space charge field which forms due to drift or diffusion [2, 4, 10, 11]. In this chapter, we shall study the modulation instability of quasi-plane-wave optical beams in biased photorefractive photovoltaic crystals under steady-state conditions. This formulation can act as a general formulation which can be reduced for the case of non-photovoltaic photorefractive crystals(where $E_{\mathrm{p}} = 0$) and photovoltaic crystals (no external bias). Also, we shall see how modulation instability affects the quasi plane wave beams when the photorefractive crystal is centrosymmetric or when the external bias is replaced by the transient pyroelectric field.

A. Katti and R. A. Yadav, *Optical Spatial Solitons in Photorefractive Materials*, Progress in Optical Science and Photonics 14, https://doi.org/10.1007/978-981-16-2550-3_3

3.2 Theoretical Foundation

3.2.1 Biased Photovoltaic Photorefractive Crystals

Consider a broad optical beam propagating in a biased photovoltaic photorefractive crystal and investigate the modulation instability of such a configuration under steady-state conditions. Consider a light beam propagating along the z axis in a photovoltaic photorefractive crystal. Further assume that the diffraction occurs along the x axis only. The photovoltaic photorefractive crystal considered in the present illustration will be $LiNbO_3$. The crystal is kept so that its c-axis is parallel to the x-axis. Also, the optical beam polarization is along the x-axis and the external electric field is applied parallel to the x-axis. The induced space charge field in case of photovoltaic photorefractive crystals considering typical photorefractive approximations is [4] (see Note 1 at the end of the chapter),

$$
\begin{aligned}
E_{sc} = & \left(E_0\frac{1}{1+|U|^2} - E_p\frac{|U|^2}{1+|U|^2} + E_p\frac{|U|^2}{1+|U|^2} f L_D \frac{\partial}{\partial x}\frac{E_{sc}}{E_t}\right) \\
& \left(1 + L_D\frac{\partial}{\partial x}\frac{E_{sc}}{E_t}\right)\left(1 - fL_D\frac{\partial}{\partial x}\frac{E_{sc}}{E_t}\right)^{-1} - \frac{K_bT}{e}\left\{\frac{\partial}{\partial x}\ln(1+|U|^2)\right. \\
& \left. - \left[\left(1 + L_D\frac{\partial}{\partial x}\frac{E_{sc}}{E_t}\right)^{-1} + f\left(1 - fL_D\frac{\partial}{\partial x}\frac{E_{sc}}{E_t}\right)^{-1}\right] L_D\frac{\partial^2}{\partial x^2}\frac{E_{sc}}{E_t}\right\}
\end{aligned}
\tag{3.1}
$$

where we assume a bright-like light beam U(x,z), i.e., $|U|^2$ tends towards zero at regions far away from the beam centre. $L_D = (\epsilon_0\epsilon_r\ k_B T/e^2 N_A)^{1/2}$ is the Debye length or the diffusion length, T is the temperature, $E_t = \frac{eN_A L_D}{\epsilon_0\epsilon_r}$, $f = {N_A}/{(N_D - N_A)}$ and $E_p = {\kappa\gamma N_A}/{e\mu}$ which is the photovoltaic field constant. μ is the electron mobility and other symbols have their usual meaning as defined before. E_0 can be obtained from the potential condition, $V_0 = -\int_{-l/2}^{l/2} E_{sc}dx + RJS$. It is approximately, $E_0 = -(V_0\chi + E_0\delta\chi - RJS\chi)$ where,

$$\chi = 1/\int_{-l/2}^{l/2}\left[1/(1+|U|^2)\right]dx$$

$$\delta = -\int_{-l/2}^{l/2}\left[|U|^2/(1+|U|^2)\right]dx$$

$$\rho = -\int_{-l/2}^{l/2}\left[\partial\ln(1+|U|^2)/\partial x\right]dx$$

Here, S is the surface area of the electrodes, l is the size of the width of the crystal between the two electrodes, R is the value of the external resistance used in the circuit, $\vec{J} = J\hat{i}$ is the total electric current density, and V_0 is the EMF of the source.

As usual, the light beam envelope follows the paraxial diffraction equation,

$$i\frac{\partial U}{\partial z} + \frac{1}{2k}\frac{\partial^2 U}{\partial x^2} - \frac{k_0}{2}n_e^3 r_{33} E_{sc} U = 0 \tag{3.2}$$

We shall consider quasi-plane wave planar waves, i.e., the amplitude $|U|$ is relatively constant over a large range of x. So $\frac{1}{2k}\frac{\partial^2 U}{\partial x^2}$ term in (3.2) can be neglected, as can the spatial derivatives of E_{sc} and $|U|^2$ in (3.1). Neglecting diffusion effects, the space charge field approximately becomes,

$$E_{sc} = \frac{E_0}{1 + |U|^2} - E_p \frac{|U|^2}{1 + |U|^2} \tag{3.3}$$

When $(2k)^{-1}\left(\partial^2 U/\partial x^2\right) = 0$, (3.2) has a solution given by,

$$U = r^{1/2} \exp\left[-i\beta\left(E_0 \frac{1}{1+r} - E_p \frac{r}{1+r}\right)\right] \tag{3.4}$$

with, $\beta = k_0 n_e^3 r_{33}/2$ and r is the peak intensity ratio, i.e., the peak intensity divided by the dark irradiance. To investigate the stability of this bright-like beam, we shall consider perturbations of the form,

$$U = \left[r^{1/2} + \sigma(x, z)\right] \exp\left[-i\beta\left(E_0 \frac{1}{1+r} - E_p \frac{r}{1+r}\right)\right] \tag{3.5}$$

where $\sigma(x, z)$ is a weak complex perturbation and its amplitude is much smaller than that of the quasi plane wave solution $|\sigma(x, z)| << r^{1/2}$,

$$\sigma(x, z) = a(z)\exp(ipx) + b(z)\exp(-ipx) \tag{3.6}$$

Putting (3.6) into (3.2) and along with (3.1), we get the following coupled equations,

$$i\frac{\partial \sigma}{\partial z} + \frac{1}{2k}\frac{\partial^2 \sigma}{\partial x^2} - \beta E\left(r^{1/2} + \sigma\right) = 0 \tag{3.7}$$

$$E - \eta\frac{\partial E}{\partial x} - \alpha\frac{\partial^2 E}{\partial x^2} = -\frac{r^{1/2}}{1+r}\left[E_{01}(\sigma + \sigma*) + \frac{k_B T}{e}\left(\frac{\partial \sigma}{\partial x} + \frac{\partial \sigma*}{\partial x}\right)\right] \tag{3.8}$$

where,

$$E = E_{sc} - E_0[1/(1+r)] + E_p[r/(1+r)]$$
$$E_{01} = \left(E_p + E_0\right)/(1+r)$$
$$\eta = \left\{E_0[1/(1+r)] - E_p[r/(1+r)]\right\}[\varepsilon_0\varepsilon_r/(eN_A)]$$
$$\alpha = [\varepsilon_0\varepsilon_r/(eN_A)](k_BT/e)$$

From (3.8) and (3.6), we can obtain the space charge field and perturbation in the spatial frequency domain by means of a fourier transform,

$$\hat{E} = -\frac{r^{1/2}}{1+r}\left\{\frac{E_{01} + E_p\alpha + i\left[E_{01}\eta k_x + k_x(k_BT/e)\left(1+\alpha k_x^2\right)\right]}{(1+\alpha k_x^2)^2 + \eta^2 k_x^2}\right\}\left(\hat{\sigma} + \hat{\sigma}*\right) \quad (3.9)$$

$$\hat{\sigma} = \int_{-\infty}^{+\infty} \exp(-ik_x x)\left[a\exp(ipx) + b\exp(-ipx)\right]dx \quad (3.10)$$

From (3.10), we have,

$$\hat{\sigma} + \hat{\sigma}* = 2\pi[(b+a*)\delta(k_x + p) + (a+b*)\delta(k_x - p)] \quad (3.11)$$

where δ signifies a delta function here. Substituting (3.11) in (3.9), we can obtain the space charge field in space by using the inverse Fourier transform,

$$E = -\frac{1}{\beta r^{1/2}}\left[G*(p)(b+a*)\exp(-ipx) + G(p)(a+b*)\exp(ipx)\right] \quad (3.12)$$

where,

$$G(p) = \beta\frac{r}{1+r}\left\{\frac{E_{01} + E_p\alpha + i\left[(E_{01}\eta + k_BT/e)p + \alpha(k_BT/e)p^3\right]}{(1+\alpha p^2)^2 + \eta^2 p^2}\right\} \quad (3.13)$$

Substituting (3.6) and (3.12) in (3.7), a set of coupled differential equations is obtained,

$$i\frac{da}{dz} - \frac{p^2}{2k}a + G(p)(a+b*) = 0 \quad (3.14)$$

$$i\frac{db}{dz} - \frac{p^2}{2k}b + G(p)(b+a*) = 0 \quad (3.15)$$

where only the linear terms have been considered.

From (3.14) and (3.15),

$$\frac{d^2a}{dz^2} = \left[\frac{p^2}{k}G(p) - \frac{p^4}{4k^2}\right]a \quad (3.16)$$

$$\frac{d^2b}{dz^2} = \left[\frac{p^2}{k}G(p) - \frac{p^4}{4k^2}\right]b \tag{3.17}$$

We can see that the solution of the Eqs. (3.16) and (3.17) is exponential in nature. For modulation instability, we need an exponential gain of the perturbation while no modulation instability implies an attenuation of the perturbation. Hence, we can find the global modulation instability gain,

$$\Gamma_{gl} = \mathrm{Re}\left\{\left[\frac{p^2}{k}G(p) - \frac{p^4}{4k^2}\right]^{1/2}\right\} \tag{3.18}$$

We can see from (18) that the maximum global MI gain occurs for low spatial-frequency. As an aside, note that there are certain photovoltaic materials which reverse the sign of their photovoltaic constants κ(i.e. E_{p}) under polarization rotation and hence the Γ_{gl} can be modified by rotating the plane of polarization.

We have used the term, "global" when describing about the modulation instability gain till now in all of the above configurations. As opposed to this, the "local" modulation instability gain can also be found out by treating the space charge field locally. This implies that we consider a broad incident light beam and neglect any diffusion effects while deriving the space charge field. In photovoltaic photorefractives, the space charge field reduces to (see Note 1 at the end of the chapter), $E_{sc} = \left(E_0\frac{1}{1+|U|^2} - E_p\frac{|U|^2}{1+|U|^2}\right)$. The ensuing procedure for deriving the local modulation instability gain remains similar to the one pursued till now by considering the weak sideband perturbation to the plane wave solution and substituting it in the dynamical evolution equation to obtain coupled differential equations for the perturbations.

3.2.2 *Biased Photorefractive Crystals*

Now, it will be in order to discuss the modulation instability in non photovoltaic photorefractive crystals, i.e., those photorefractives which have a negligible photovoltaic constant. This implies $E_{\mathrm{p}} = 0$. For a short circuit, we also have $R = 0$ and hence, $E_0 = E_{0S} = -V_0\chi$. So from (3.13) and (3.18),

$$G_S(p) = \beta\frac{r}{1+r}\left\{\frac{E_s + i\left[(E_S\eta_S + k_BT/e)p + \alpha(k_BT/e)p^3\right]}{(1+\alpha p^2)^2 + \eta_S^2p^2}\right\} \tag{3.19}$$

$$\Gamma_{gl} = \mathrm{Re}\left\{\left[\frac{p^2}{k}G_S(p) - \frac{p^4}{4k^2}\right]^{1/2}\right\} \tag{3.20}$$

where, $E_S = \frac{E_{0S}}{1+r}$ and $\eta_S = E_{0S}[1/(1+r)][\epsilon_0\epsilon_r/(eN_A)]$

3.2.3 Unbiased Photovoltaic Photorefractive Crystals

If the external bias is not present, this reduces to the case of either a closed circuit photovoltaic crystal ($J \neq 0$) or an open circuit photovoltaic crystal($J = 0$). So, the total electric current density can be represented as [12], $J_{p1} = \frac{E_{p0}}{E_p}$ where $J_{p1} = \frac{J}{(\hat{s} I_d N_D \kappa)}$ and $E_{p0} = -\big(E_p \delta \chi - RJS\chi\big)$ where $\hat{s}$ is the photoexcitation cross section and other symbols have their usual meanings. If there is no external electric field or voltage bias, then $E_0 = E_{p0}$. So from (3.13) and (3.18),

$$G(p) = \beta \frac{r}{1+r} \left\{ \frac{E_{p1} + i\big[\big(E_{p1}\eta_p + k_B T/e\big)p + \alpha(k_B T/e)p^3\big]}{(1+\alpha p^2)^2 + \eta_p^2 p^2} \right\} \tag{3.21}$$

$$\Gamma_{gl} = \mathrm{Re}\left\{ \left[\frac{p^2}{k} G_p(p) - \frac{p^4}{4k^2} \right]^{1/2} \right\} \tag{3.22}$$

where, $E_{p1} = E_p\big(J_{p1} + 1\big)/(1+r)$ and $\eta_p = E_p\big(J_{p1} - r\big)[1/(1+r)][\epsilon_0 \epsilon_r/(eN_A)]$.

3.2.4 Centrosymmetric Photorefractive Crystals

In photorefractive crystals with an inversion symmetry, i.e., centrosymmetric photorefractive crystals, the nonlinearity is due to the quadratic electro-optic effect and hence the refractive index change is,

$$\Delta n = -\frac{1}{2} n_e^3 g_{eff} \, \epsilon_0^2 \, (\epsilon_r - 1)^2 E_{sc}^2 / 2 \tag{3.23}$$

Using (3.23), the light beam envelope follows the paraxial diffraction equation,

$$i \frac{\partial U}{\partial z} + \frac{1}{2k} \frac{\partial^2 U}{\partial x^2} - \frac{k_0}{2} n_e^3 g_{eff} \, \epsilon_0^2 \, (\epsilon_r - 1)^2 E_{sc}^2 U = 0 \tag{3.24}$$

We shall consider quasi-plane wave planar beams, i.e., the amplitude $|U|$ is relatively constant over a large range of x. So $\frac{1}{2k}\frac{\partial^2 U}{\partial x^2}$ term in (3.24) can be neglected, as can the spatial derivatives of E_{sc} and $|U|^2$ in (1). Without any approximations, the global space charge field can be expressed (see (1.43) of Chap. 1),

$$E_{sc} = E_0 \frac{1}{1+|U|^2} \left(1 + \frac{\epsilon_0 \epsilon_r}{eN_A} \frac{\partial E_{sc}}{\partial x} \right) - \frac{k_B T}{e} \frac{(\partial |U|^2 / \partial x)}{(I + I_d)} + \frac{k_B T}{e} \frac{\epsilon_0 \epsilon_r}{eN_A} \left(1 + \frac{\epsilon_0 \epsilon_r}{eN_A} \frac{\partial E_{sc}}{\partial x} \right)^{-1} \frac{\partial^2 E_{sc}}{\partial x^2} \tag{3.25}$$

when $(2k)^{-1}\left(\partial^2 U/\partial x^2\right) = 0$, (3.24) has a solution given by,

$$U = r^{1/2}\exp\left[-i\beta\left(\frac{E_0^2}{(1+r)^2}\right)z\right] \tag{3.26}$$

with, $\beta = k_0 n_e^3 g_{eff} \in_0^2 (\in_r - 1)^2/2$. For investigating the stability of this bright-like beam, we shall consider perturbations of the form,

$$U = \left(r^{\frac{1}{2}} + \sigma(x, z)\right)\exp\left[-i\beta\left(\frac{E_0^2}{(1+r)^2}\right)z\right] \tag{3.27}$$

where $\sigma(x, z)$ is a weak complex perturbation and its amplitude is much smaller than that of the quasi plane wave solution $|\sigma(x, z)| << r^{1/2}$,

$$\sigma(x, z) = a(z)\exp(ipx) + b(z)\exp(-ipx) \tag{3.28}$$

Putting (3.28) into (3.24) along with (3.27), the following coupled equations are obtained,

$$i\frac{\partial\sigma}{\partial z} + \frac{1}{2k}\frac{\partial^2\sigma}{\partial x^2} - \beta E(E + 2E_{01})\left(r^{1/2} + \sigma\right) = 0 \tag{3.29}$$

$$E - \alpha\frac{\partial E}{\partial x} - \mu\frac{\partial^2 E}{\partial x^2} = -\frac{r^{1/2}}{1+r}\left[E_{01}(\sigma + \sigma*) + \frac{k_B T}{e}\left(\frac{\partial\sigma}{\partial x} + \frac{\partial\sigma*}{\partial x}\right)\right] \tag{3.30}$$

where,
$$E = E_{sc} - E_{01}$$
$$E_{01} = E_0/(1+r)$$
$$\alpha = E_{01}[\varepsilon_0\varepsilon_r/(eN_A)]$$
$$\mu = (k_B T/e)[\varepsilon_0\varepsilon_r/(eN_A)]$$

From (3.30) and (3.28), we can obtain the space charge field and perturbation in the spatial frequency domain by means of a fourier transform,

$$\hat{E} = -\frac{r^{1/2}}{1+r}\left\{\frac{E_{01} + i\left[E_{01}\alpha k_x + k_x(k_B T/e)\left(1 + \mu k_x^2\right)\right]}{(1 + \mu k_x^2)^2 + \alpha^2 k_x^2}\right\}\left(\hat{\sigma} + \hat{\sigma}*\right) \tag{3.31}$$

$$\hat{\sigma} = \int_{-\infty}^{+\infty}\left[a\exp(ipx) + b\exp(-ipx)\right]\exp(-ik_x x)dx \tag{3.32}$$

From (3.32), we have,

$$\hat{\sigma} + \hat{\sigma}* = 2\pi[(b + a*)\delta(k_x + p) + (a + b*)\delta(k_x - p)] \tag{3.33}$$

δ signifies a delta function in (3.33). Putting (3.33) into (3.31) and using the inverse Fourier transform to obtain the induced space charge field in space,

$$E = -\frac{1}{\beta r^{1/2}}\left[G*(p)(b + a*)\exp(-ipx) + G(p)(a + b*)\exp(ipx)\right] \tag{3.34}$$

where,

$$G(p) = \beta\frac{r}{1+r}\left\{\frac{E_{01} + i\left[(E_{01}\alpha + k_B T/e)p + \mu(k_B T/e)p^3\right]}{(1+\mu p^2)^2 + \alpha^2 p^2}\right\} \tag{3.35}$$

Substituting (3.28) and (3.34) in (3.29), the following coupled differential equations are obtained in the linear approximation,

$$i\frac{da}{dz} - \frac{p^2}{2k}a + 2E_{01}G(p)(a + b*) = 0 \tag{3.36}$$

$$i\frac{db}{dz} - \frac{p^2}{2k}b + 2E_{01}G(p)(b + a*) = 0 \tag{3.37}$$

From (3.36) and (3.37),

$$\frac{d^2a}{dz^2} = \left[\frac{2E_{01}p^2}{k}G(p) - \frac{p^4}{4k^2}\right]a \tag{3.38}$$

$$\frac{d^2b}{dz^2} = \left[\frac{2E_{01}p^2}{k}G(p) - \frac{p^4}{4k^2}\right]b \tag{3.39}$$

We can see that the (3.38) and (3.39) have exponential solutions. For modulation instability, an exponential gain of the perturbation is required while no modulation instability needs an attenuation of the perturbation. Hence, the global modulation instability gain is,

$$\Gamma_{gl} = \mathrm{Re}\left\{\left[\frac{2E_{01}p^2}{k}G(p) - \frac{p^4}{4k^2}\right]^{1/2}\right\} \tag{3.40}$$

We have used the term, "global" when describing about the modulation instability gain till now in all of the above configurations. As opposed to this, the "local" modulation instability gain can also be found out by treating the space charge field locally. This implies that we consider a broad incident light beam and neglect any diffusion effects while deriving the space charge field. In centrosymmetric photorefractives, the space charge field reduces to, $E_{sc} = E_0\frac{1}{1+|U|^2}$

The ensuing procedure for deriving the local modulation instability gain remains similar to the one pursued till now by considering the weak sideband perturbation to the plane wave solution and substituting it in the dynamical evolution to obtain coupled differential equations for the perturbations. Figure 3.1 shows the global and

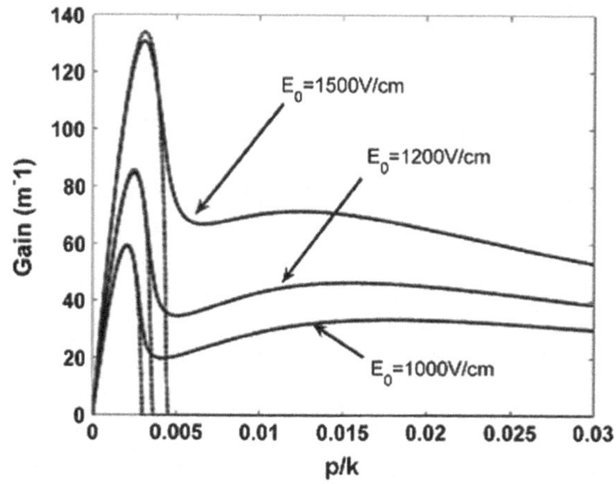

Fig. 3.1 Global modulation instability gain(shown by the solid line) and local modulation gain(shown by the dashed lines) versus *p/k*, $r = 2$ in a centrosymmetric photorefractive crystal (Reprinted from Physics Letters A, 374, Kaiyun Zhan and Chunfeng Hou, One-dimensional modulational instability of broad optical beams in biased centrosymmetric photorefractive crystals, 169–172, Copyright 2009, with permission from Elsevier)

Table. 3.1 Parameters of KLTN crystal taken in the present analysis [10]

n_e	2.2	T	294 K
N_A	10^{22} /m^3	β	4.03×10^{-8} /V^2
g_{eff}	0.12 m^4/C^2	k	2.76×10^7 m^{-1}
ε_r	8000	λ_0	0.5 μm

local modulation instability gain in case of centrosymmetric photorefractives taking KLTN crystal parameters as in Table 3.1.

3.2.5 *Pyroelectric Photorefractive Crystals*

In this case, we consider a photorefractive crystal exhibiting the pyroelectric effect. The refractive index change is now because of the pyroelectric space charge field solely (see Section 2.4). The crystal has its temperature controlled by a metallic plate in contact with it which is in turn connected to an external agency. An insulating plastic cover is used to minimize any external influence on the crystal. Rest of the configuration of the optical beam and diffraction compensation is similar to the previous considerations. Strontium Barium Niobate, $Sr_{0.6}Ba_{0.4}Nb_2O_6$ (SBN) henceforth can be considered as an excellent example for illustration [13, 14].

E_{sc} is the induced space charge field which is now formed solely due to the transient pyroelectric field. Hence $E_{sc} = E_{pysc}$ and is [13],

$$E_{pysc} = -E_{py}\frac{I}{I+I_d} \tag{3.41}$$

E_{py} is the transient pyroelectric field and is given by [15],

$$E_{py} = -\frac{1}{\varepsilon_0\varepsilon_r}\frac{\partial P}{\partial T}\Delta T \tag{3.42}$$

where $\frac{\partial P}{\partial T}$ is the pyroelectric coefficient. The magnitude of the temperature change of the crystal is given by ΔT. Other symbols have their usual meanings. The value and sign of E_{py} can be changed by adjusting the change in temperature, and so this implies a strong correlation between the magnitude of heating or cooling and self trapping including the modulation instability as we shall see further. Using (3.42), we can easily obtain the following equation of dynamical evolution using the usual dimensionless coordinates,

$$iU_z + \frac{1}{2}U_{xx} + \beta E_{py}\frac{|U|^2}{1+|U|^2}U = 0 \tag{3.43}$$

For studying the modulation instability, we need to take a plane wave broad beam solution of (3.43),

$$U = r^{1/2}\exp\left[i\beta E_{py}\left\{\frac{r}{1+r}\right\}z\right] \tag{3.44}$$

and express the perturbed solution as,

$$U = \left[r^{1/2} + \sigma(x,z)\right]\exp\left[i\beta E_{py}\left\{\frac{r}{1+r}\right\}z\right] \tag{3.45}$$

$\sigma(x,z)$ is a weak modulation term added to the steady state solution containing two sideband plane waves. This weak perturbation satisfies

$$|\sigma(x,z)|^2 << r^{1/2} \tag{3.46}$$

The form of the weak perturbation can be expressed,

$$\sigma = a(z)\exp(ipx) + b(z)\exp(-ipx) \tag{3.47}$$

The next step is to check whether this perturbation grows exponentially or is attenuated with propagation. Hence, substituting (3.47) in (3.45) along with (3.46), we obtain the evolution equation satisfied by the perturbation $\sigma(x,z)$,

$$i\frac{\partial\sigma}{\partial z} + \frac{1}{2k}\frac{\partial^2\sigma}{\partial x^2} + \beta E_{py}\frac{r}{(1+r)^2}(\sigma+\sigma*) = 0 \tag{3.48}$$

Substituting (3.47) in (3.48), we obtain,

$$i\frac{da}{dz} - \frac{1}{2k}p^2a + \beta E_{py}\frac{r}{(1+r)^2}(a+b*) \tag{3.49}$$

$$i\frac{db}{dz} - \frac{1}{2k}p^2b + \beta E_{py}\frac{r}{(1+r)^2}(a*+b) \tag{3.50}$$

Decoupling Eqs. (3.49) and (3.50) , we get,

$$\frac{d^2a}{dz^2} = \left[\beta E_{py}\frac{r}{(1+r)^2}\frac{p^2}{k} - \frac{p^4}{4k^2}\right]a \tag{3.51}$$

$$\frac{d^2b}{dz^2} = \left[\beta E_{py}\frac{r}{(1+r)^2}\frac{p^2}{k} - \frac{p^4}{4k^2}\right]b \tag{3.52}$$

Equations (3.51) and (3.52) are coupled differential equations. Therefore, their general solution can be expressed as an exponential ~ $\exp(\omega\xi)$ with

$$\omega = \left[\beta E_{py}\frac{r}{(1+r)^2}\frac{p^2}{k} - \frac{p^4}{4k^2}\right]^{1/2} \tag{3.53}$$

The local modulation instability gain can be found from (3.53),

$$g = \mathrm{Re}\left\{\left[\beta E_{py}\frac{r}{(1+r)^2}\frac{p^2}{k} - \frac{p^4}{4k^2}\right]^{1/2}\right\} \tag{3.54}$$

Also, the maximum modulation instability gain will be,

$$g_{\max} = \left[\frac{1}{2}k_0 E_{py} n_e^3 r_{eff}\frac{r}{(1+r)^2}\right] \tag{3.55}$$

and the spatial frequency associated with the maximum modulation instability gain is,

$$p_{max} = \frac{k_0 n_e^2}{1+r}\left[r_{eff}E_{py}r\right]^{1/2} \tag{3.56}$$

Considering the SBN crystal's parameters [13, 16, 17], $n_e = 2.35$, $\lambda_0 = 532$ nm., $g_{\text{eff}} = 237 \times 10^{-12}$ m/V, $\varepsilon_0 = 8.85 \times 10^{-12}$ F/m, $\varepsilon_r = 3400$, $\frac{\partial P}{\partial T} = -3 \times 10^{-4}$ $\mathrm{Cm^{-2}\,K^{-1}}$, $r = 10$. Using the parameters elucidated above, $\beta = 0.018$ and $k = 2.774 \times 10^7$. The MI gain's variation with *p/k* at a particular temperature is shown in Fig. 3.2. *p/k* is the angle at which the plane wave components of the $\sigma(x, z)$ perturbation propagate with respect to the broad optical beam.

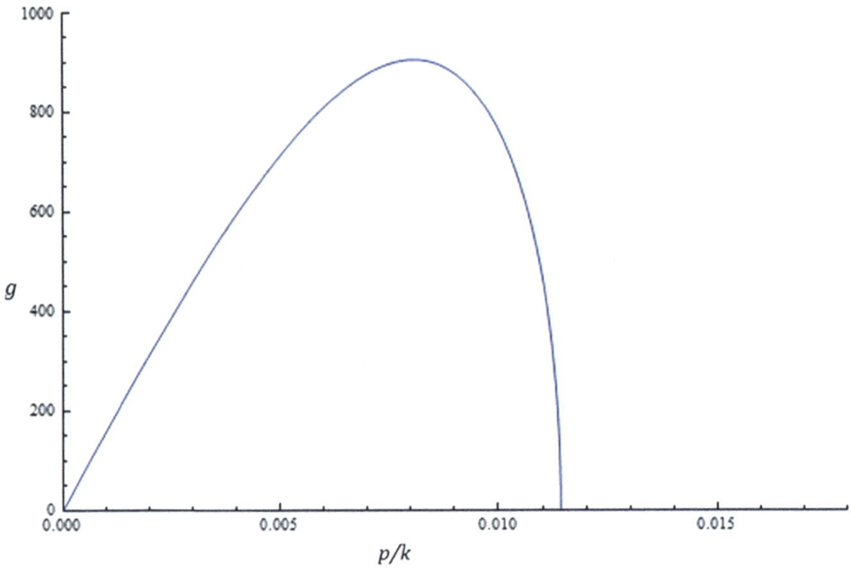

Fig. 3.2 Modulation Instability gain g as a function of p/k with $\Delta T = 20\,°\mathrm{C}$. (Reprinted from Chaos, Solitons and Fractals, 101,Aavishkar Katti and R.A.Yadav, Modulation instability of broad optical beams in unbiasedphotorefractive pyroelectric crystals, 20–23, Copyright 2017, with permission from Elsevier)

From (3.54) and (3.55), the modulation instability gain is a function of the temperature change of the photorefractive pyroelectric crystal. Figure 3.3 shows the simultaneous plot showing the variation of the modulation instability gain g with ΔT and p/k for $r = 1$. As the value of ΔT increases, the peak modulation instability gain g_{max} increases. Again, as the value of ΔT increases, the range of values of p/k corresponding to a finite gain increases with a rise in ΔT.

3.3 Concluding Remarks and Further Reading

A nice introduction to modulation instability and its beginning is given in [1]. As in the case of solitons, modulation instability was first observed in water waves [7, 8]. The same concept was then found to be true for optical light beams. The reader is referred to [2, 4, 9, 16–18] for a detailed study of modulation instability in different types and configurations of photorefractive crystals.

Note 1:

The induced space charge field can be derived from the set of rate equations, continuity equations and Gauss Law in one dimension for steady state [18],

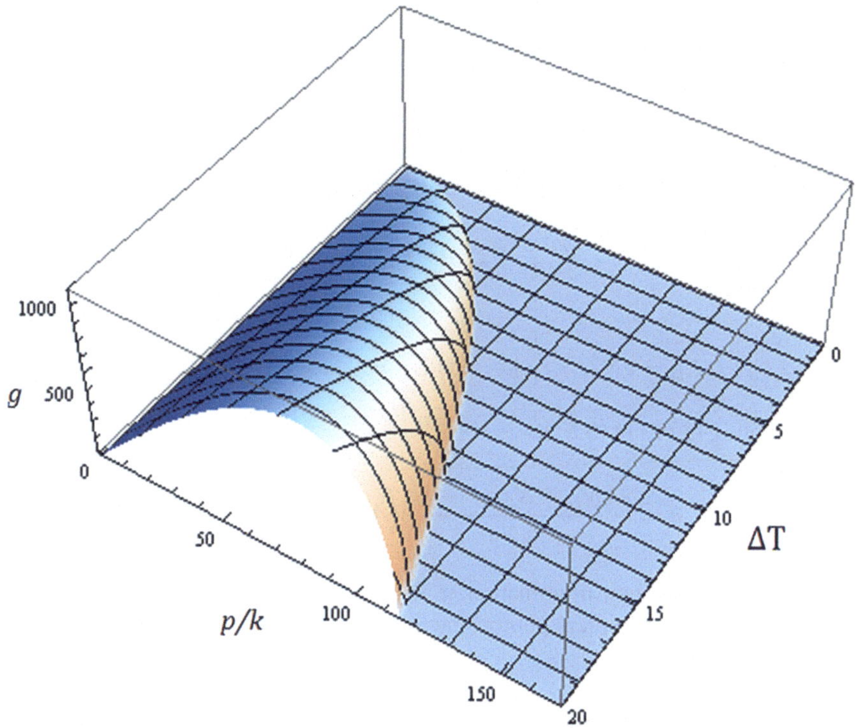

Fig. 3.3 Modulation Instability gain with respect to *p/k* and ΔT when $r = 1$. (*p/k* is scaled in units of 10^4). (Reprinted from Chaos, Solitons and Fractals, 101,Aavishkar Katti and R.A.Yadav, Modulation instability of broad optical beams in unbiasedphotorefractive pyroelectric crystals, 20–23, Copyright 2017, with permission from Elsevier)

$$\gamma_R n N_D^+ = s_i(I + I_d)\left(N_D - N_D^+\right) \tag{3.57}$$

$$\frac{\partial E_{sc}}{\partial x} = \frac{e}{\epsilon_0 \epsilon}\left(N_D^+ - N_A - n\right) \tag{3.58}$$

$$J = e\mu n E_{sc} + k_B T \mu \frac{\partial n}{\partial x} + k_p s_i\left(N_D - N_D^+\right) I \tag{3.59}$$

$$\frac{\partial J}{\partial x} = 0 \tag{3.60}$$

where the symbols have their usual meanings. There is one change in these equations if we compare them with those used in previous chapters. We have now considered k_p to be the photovoltaic constant which contributes the photovoltaic current term. As usual, any z spatial dependence has been ignored assuming a much more rapid variation in x. Now, in typical photovoltaic-photorefractive media,$N_D^+ >> n$,$N_D >>$

n, and $N_A >> n$. Hence, (3.57) and (3.58) give,

$$N_D^+ = N_A\left(1 + \frac{\epsilon_0\epsilon_r}{eN_A}\frac{\partial E_{sc}}{\partial x}\right) \tag{3.61}$$

(3.61) can be represented as,

$$N_D^+ = N_A\left(1 + L_D\frac{\partial}{\partial x}\frac{E_{sc}}{E_t}\right) \tag{3.62}$$

where,

$E_t = k_BT/eL_D = eN_AL_D/\,\epsilon_0\epsilon_r$ and $L_D = (\epsilon_0\epsilon_r\, k_BT/e^2N_A)^{1/2}$ is the Debye length.

Substituting (3.62) in (3.57),

$$n = \frac{s_i(N_D - N_A)(I + I_d)}{\gamma_R N_A}\left(1 + L_D\frac{\partial}{\partial x}\frac{E_{sc}}{E_t}\right)^{-1} \tag{3.63}$$

In regions of constant illumination, $x \to \pm\infty$, $E(x \to \pm\infty) = E_0$(constant), $\partial E/\partial x = 0$. Also, in typical photorefractive materials, $N_D^+ = N_A$. Hence, (3.59) and (3.63) yield,

$$J_\infty = J(x \to \pm\infty, z) = e\mu n_\infty E_0 + k_p s_i(N_D - N_A)I_\infty \tag{3.64}$$

With

$$n_\infty = n(x \to \pm\infty) = \frac{s_i(N_D - N_A)(I + I_d)}{\gamma_R N_A} \tag{3.65}$$

From equation, we can infer that the current is constant everywhere, $J = J_\infty$ and so from (3.59) and (3.64) using (3.63) and (3.65),

$$\begin{aligned} E_{sc} = {} & \left(E_0\frac{1}{1+|U|^2} - E_p\frac{|U|^2}{1+|U|^2} + E_p\frac{|U|^2}{1+|U|^2} f L_D\frac{\partial}{\partial x}\frac{E_{sc}}{E_t}\right) \\ & \left(1 + L_D\frac{\partial}{\partial x}\frac{E_{sc}}{E_t}\right)\left(1 - fL_D\frac{\partial}{\partial x}\frac{E_{sc}}{E_t}\right)^{-1} - \frac{K_bT}{e}\left\{\frac{\partial}{\partial x}\ln(1+|U|^2)\right. \\ & \left. - \left[\left(1 + L_D\frac{\partial}{\partial x}\frac{E_{sc}}{E_t}\right)^{-1} + f\left(1 - fL_D\frac{\partial}{\partial x}\frac{E_{sc}}{E_t}\right)^{-1}\right] L_D\frac{\partial^2}{\partial x^2}\frac{E_{sc}}{E_t}\right\} \end{aligned} \tag{3.66}$$

$L_D = (\epsilon_0\epsilon_r\, k_BT/e^2N_A)^{1/2}$ is the Debye length or the diffusion length, T is the temperature, $E_t = \frac{eN_AL_D}{\epsilon_0\epsilon_r}$, $f = {}^{N_A}\!/_{(N_D - N_A)}$ and $E_p = {}^{\kappa\gamma N_A}\!/_{e\mu}$ which is the photovoltaic field constant. μ is the electron mobility and other symbols have their usual meaning as defined before.

If the intensity of the light beam varies relatively slowly with respect to x, the term $L_D \frac{\partial E_{sc}}{\partial x}$ can be ignored, as it is of the order of much less than unity. Also, the diffusion terms can be neglected if the external electric field's magnitude is moderately large and the drift current terms dominate over the diffusion terms. Finally, we can obtain the space charge field from (3.66) as,

$$E_{sc} = E_0 \frac{I_\infty + I_d}{I + I_d} + E_p \frac{I_\infty - I}{I + I_d} - \frac{k_B T}{e} \frac{1}{I + I_d} \frac{\partial I}{\partial x} \tag{3.67}$$

References

1. V.E. Zakharov, L.A. Ostrovsky, Modulation instability: the beginning. Physica D **238**(5), 540–548 (2009). https://doi.org/10.1016/j.physd.2008.12.002
2. M.I. Carvalho, S.R. Singh, D.N. Christodoulides, Modulational instability of quasi-plane-wave optical beams biased in photorefractive crystals. Optics Commun. **126**(1–3), 167–174 (1996). https://doi.org/10.1016/0030-4018(95)00743-1
3. N.N. Akhmediev, V.I. Korneev, Modulation instability and periodic solutions of the nonlinear Schr?? dinger equation. Theor. Math. Phys. **69**(2), 1089–1093 (1986). https://doi.org/10.1007/BF01037866
4. L. Ke-Qing, Z. Wei, Y. Yan-Long, Modulation instability in biased photorefractive-photovoltaic crystals. Chin. Phys. **21**(6), 1086–1088 (2004)
5. D. Kip, Modulation instability and pattern formation in spatially incoherent light beams. Science **290**(5491), 495–498 (2000). https://doi.org/10.1126/science.290.5491.495
6. S. Wen, Y. Wang, W. Su, Y. Xiang, X. Fu, D. Fan, Modulation instability in nonlinear negative-index material. Phys. Rev. E—Stat. Nonlinear Soft Matter Phys. **73**(3), 036617 (2006). https://doi.org/10.1103/PhysRevE.73.036617
7. T.B. Benjamin, J.E. Feir, The disintegration of wave trains on deep water Part 1. Theory. J. Fluid Mech. **27**(3), 417–430 (1967)
8. J.W. McLean, Instabilities of finite-amplitude water waves. J. Fluid Mech. **114**, 315–330 (1982)
9. K.Q. Lu, W. Zhao, Y.L. Yang, X.P. Zhu, J.P. Li, Y.P. Zhang, Modulation instability of quasi-plane-wave optical beams in biased photorefractive-photovoltaic crystals. Chin. Phys. **13**(12), 2077–2081 (2004). https://doi.org/10.1088/1009-1963/13/12/017
10. Zhan, C. Hou, "One-dimensional modulational instability of broad optical beams in biased centrosymmetric photorefractive crystals," Phys. Lett. Sect. A: Gen. At. Solid State Phys. **374**(2), 169–172 (2009), https://doi.org/10.1016/j.physleta.2009.10.037
11. A. Katti, R.A. Yadav, Modulation instability of broad optical beams in unbiased photorefractive pyroelectric crystals. Chaos, Solitons Fractals **101**, 20–23 (2017). https://doi.org/10.1016/j.chaos.2017.05.008
12. L. Keqing, Z. Yanpeng, T. Tiantong, L. Bo, "Incoherently coupled steady-state soliton pairs in biased photorefractive-photovoltaic materials," Phys. Rev. E—Stat. Nonlinear Soft Matt. Phys. **64**(5 II), 056603/1–056603/9 (2001), https://doi.org/10.1103/PhysRevE.64.056603.
13. Y. Su, Q. Jiang, X. Ji, Photorefractive spatial solitons supported by pyroelectric effects in strontium barium niobate crystals. Optik **126**(18), 1621–1624 (2015). https://doi.org/10.1016/j.ijleo.2015.04.053
14. A. Katti, R.A.A. Yadav, Incoherently coupled photorefractive spatial solitons supported by pyroelectric effects. J. Nonlinear Opt. Phys. Mater. **26**(01), 1750002 (2017). https://doi.org/10.1142/S0218863517500023
15. J. Safioui, F. Devaux, M. Chauvet, Pyroliton: pyroelectric spatial soliton. Opt. Express **17**(24), 22209–22216 (2009). https://doi.org/10.1364/OE.17.022209

16. A.A. Savchenkov, A.B. Matsko, V.S. Ilchenko, I. Solomatine, D. Seidel, L. Maleki, "RF-induced change of optical refractive index in strontium barium niobate," Photonics West, Laser Resonators, Microresonators, and Beam Control XV (February 22, 2013), vol. 8600, pp. 86000O-86000O–9, 2013, doi: https://doi.org/10.1117/12.2008511.
17. K. Kos, H. Meng, G. Salamo, M. Shih, M. Segev, G.C. Valley, One-dimensional steady-state photorefractive screening solitons. Phys Rev E Stat Phys Plasmas Fluids Relat Interdiscip Topics **53**(5), R4330–R4333 (1996). https://doi.org/10.1103/PhysRevE.53.R4330
18. N.V. Kukhtarev, V.B. Markov, S.G. Odulov, M.S. Soskin, V.L. Vinetskii, Holographic storage in electrooptic crystals. i. steady state. Ferroelectrics **22**(1), 949–960 (1978). https://doi.org/10.1080/00150197908239450

Chapter 4
Formation of Photorefractive Solitons

4.1 Introduction

The photorefractive nonlinearity has many advantages as discussed before like the saturable nature of its nonlinearity and the easy realization in experiments due to low laser power requirement. But one of the prime drawbacks of the photorefractive response is its very slow nature. The photorefractive response time can be said to be approximately equivalent to the dielectric response time. Uptill now, many investigations have been carried out for studying the formation characteristics and temporal properties of photorefractive solitons in various different types of photorefractive crystals, i.e., conventional photorefractive media [1–4], photovoltaic media [5–7], photorefractive-photovoltaic media [8], centrosymmetric photorefractive media [9], photorefractive semiconductors [10, 11], and novel photorefractive crystals exhibiting both the linear and quadratic electro-optic effect [12].

In this chapter, we shall discuss a general theory for characterising the temporal response of photorefractive solitons. The time dependent dynamical evolution equation for photorefractive solitons will be derived. The temporal evolution of the solitons can be inferred by studying the change in soliton width with respect to time for different intensity ratios. The conditions under which quasi steady state solitons form have been investigated. The dependence of temporal evolution characteristics of the solitons on magnitude of the electro-optic coefficients is studied. We then illustrate how this reduces to studying the temporal characteristics of photorefractive solitons in noncentrosymmetric and centrosymmetric crystals. Finally, the temporal evolution of photovoltaic solitons and screening photovoltaic solitons is also discussed on the basis of the previous analysis.

A. Katti and R. A. Yadav, *Optical Spatial Solitons in Photorefractive Materials*, Progress in Optical Science and Photonics 14, https://doi.org/10.1007/978-981-16-2550-3_4

4.2 Theoretical Formulation: Non-photovoltaic Photorefractive Crystals

Consider an optical beam propagating in a photorefractive crystal along the z-axis. We shall assume that the diffraction is along the x-direction only. The crystal is oriented in such a way so that its c-axis coincides with the x-axis. The polarization of the light beam is taken in the x-direction and the external electric field is also applied along x-axis. In the slowly varying envelope approximation, the incident beam's electric field can be expressed as, $E = \hat{x}A(x, z)\exp(ikz)$ where $k = k_0 n_e = (2\pi/\lambda_0)n_e$. λ_0 is the wavelength in free space, n_e is the unchanged refractive index. Hence, the dynamical evolution equation becomes,

$$\left(i\frac{\partial}{\partial z} + \frac{1}{2k}\frac{\partial^2}{\partial x^2} - b\frac{k_0 n_e^3 g_{eff} \epsilon_0^2 (\epsilon_r - 1)^2 {E_s}^2}{2} - a\frac{k_0 n_e^3 r_{eff} E_s}{2}\right)A(x, z) = 0 \tag{4.1}$$

where E_s is the induced space charge field, r_{eff} and g_{eff} denote the linear and quadratic electro-optic coefficient. $a = 0$, $b = 1$ represents the propagation equation for a centrosymmetric photorefractive crystal, $a = 1$ and $b = 0$ represents the propagation equation for a conventional photorefractive crystal exhibiting the linear electro-optic effect, while $a = 1, b = 1$ represents the propagation equation in novel photorefractive crystals exhibiting the linear and quadratic electro-optic effect simultaneously.

ϵ_0 and ϵ_r are the vacuum permittivity and relative dielectric constant. The charge transport model of Kukhtarev's charge transport model [13] will serve as the starting point to derive the space charge field,

$$\frac{\partial}{\partial t}N_D^+ = (s_i I + \beta)(N_D - N_D^+) - \gamma n N_D^+ \tag{2a}$$

$$\frac{\partial}{\partial x}(\epsilon_0 \epsilon_r E) = \rho_i \tag{2b}$$

$$\frac{\partial J}{\partial x} + \frac{\partial \rho_i}{\partial t} = 0 \tag{2c}$$

$$\rho_i = e(N_D^+ - n - N_A) \tag{2d}$$

$$J = e\mu n E + k_B T \mu \frac{dn}{dx} \tag{2e}$$

where the beam intensity is $I = |A|^2$, β is the generation rate of thermally generated or dark carriers, s_i is the photoionization cross section, the recombination rate of carriers is denoted by γ while the mobility of the electron is denoted by μ and e denotes the electron's charge. ρ_i is the total charge density while k_B represents the

Boltzmann's constant. N_A and N_D are the concentrations of acceptors and donors respectively. n is the electron density, T is the temperature and N_D^+ is the ionized donor density.

In principle, (4.2a)–(4.2e) can be solved simultaneously to obtain a time dependent space charge field. In practice, this is not so easy so a few approximations must be made to simplify the charge transport equations as follows. Firstly, $N_D^+ \approx N_A$ in typical photorefractive materials. This is so because the electron density is very small as compared to the ionized donor or acceptor density for moderate incident intensity. Secondly, the carrier recombination time is negligible with respect to the dielectric response time which gives us $\partial N_D^+/\partial t = 0$ [1]. Using these relations in (4.2a), we get,

$$(s_i I + \beta)(N_D - N_A) = \gamma n N_A \tag{3a}$$

We can easily obtain the following differential equation connecting the space charge field and the intensity,

$$e\mu\frac{\partial}{\partial x}[(I + I_d)E] + k_B T\mu\frac{\partial^2 I}{\partial x^2} + \epsilon_0\epsilon_r\frac{\gamma N_A}{s_i(N_D - N_A)}\frac{\partial^2 E}{\partial t\partial x} = 0 \tag{3b}$$

where $I_d = \beta/s_i$ is the dark irradiance. For obtaining (4.3b), substitute for ρ_i from (4.2b), J from (4.2e) in the continuity Eq. (4.2c) with n from (4.3a).

Integrating (4.3b) once, we get,

$$E + \frac{k_b T}{e(I + I_d)}\frac{\partial I}{\partial x} + \frac{C I_d}{(I + I_d)}\frac{\partial E}{\partial t} = E_0\frac{I_\infty + I_d}{I + I_d} \tag{4.4}$$

where we have applied the condition [14], $E(t \to \infty) = \frac{E_0(I_\infty + I_d)}{(I + I_d)} - \frac{k_B T}{e}\frac{(\partial I/\partial x)}{(I + I_d)}$ and $C = (\epsilon_0\epsilon_r\gamma N_A)/[e\mu\beta(N_D - N_A)]$, $I_\infty = I(x \to \pm\infty, z)$ represents the intensity in regions far away from the incident beam's center, i.e., constant illumination regions. $E_0 = E(x \to \pm\infty, z)$ is the limiting value of the induced space charge field far from the beam center, i.e., in constant illumination regions. E_0 is approximately equal to V_0/W if the spatial width of the beam is much smaller than the x-width of the photorefractive crystal. Supposing that the intensity changes slowly with progression of time [1, 4, 5, 15] and remembering the fact that $E(t \to 0) = E_0$, we obtain, from (4.4),

$$\begin{aligned} E = {} & E_0\exp\left[-\frac{(I + I_d)t}{C I_d}\right] \\ & + E_0\left(\frac{(I_\infty + I_d)}{I + I_d} - \frac{k_B T}{e(I + I_d)}\frac{\partial I}{\partial x}\right)\left\{1 - \exp\left[-\frac{(I + I_d)t}{C I_d}\right]\right\} \end{aligned} \tag{4.5}$$

The $(k_B T/e)$ terms or the diffusion field terms can be ignored for a relatively strong external bias. Hence (4.5) becomes,

Table 4.1 Parameters considered for PMN-0.33PT crystal [16–20]

Parameter	Value	Parameter	Value
λ_0	632.8 nm	V	1000 V
x_0	20 μm	l *(crystal width)*	1 cm
n_e	2.562	β_1	15.4612
r_{eff}	182×10^{-12} m/V	β_2	1.1724
$g_{eff}\epsilon_0^2(\epsilon_r - 1)^2$	$1.38 \times 10^{-16}\ \mathrm{m^2/V^2}$	ρ	0

$$E = E_0\left[\frac{(I_\infty + I_d)}{I + I_d}\right] + E_0\left(\frac{(I - I_\infty)}{I + I_d}\right)\left\{\exp\left[-\frac{(I + I_d)t}{CI_d}\right]\right\} \tag{4.6}$$

Substitute (4.6) in (4.1) to obtain dynamical evolution equation as an explicit function of time,

$$\begin{aligned} iU_\xi + \frac{1}{2}U_{ss} - \beta_2\frac{\left\{1 + \rho + \left(|U|^2 - \rho\right)\exp\left[-\left(1 + |U|^2\right)\tau\right]\right\}^2}{\left(1 + |U|^2\right)^2}U \\ - \beta_1\frac{\left\{1 + \rho + \left(|U|^2 - \rho\right)\exp\left[-\left(1 + |U|^2\right)\tau\right]\right\}}{\left(1 + |U|^2\right)}U = 0 \end{aligned} \tag{4.7}$$

where, $\beta_1 = a\frac{(k_0x_0)^2 n_e^4 r_{eff}}{2}E_0$, $\beta_2 = b\frac{(k_0x_0)^2 n_e^4 g_{eff}\epsilon_0^2(\epsilon_r - 1)^2}{2}E_0^2$, $\rho = I_\infty/I_d$ and the following dimensionless coordinates have been used,

$\tau = t/C$, $s = x/x_0$, $\xi = z/kx_0^2$, $A = (2\eta_0 I_d/n_e)^{1/2}U$ where, $\eta_0 = (\mu_0/\epsilon_0)^{1/2}$ and x_0 is the scale parameter. For illustration of the results, consider the parameters for a PMN-0.33PT crystal as given in Table 4.1. Notably such a photorefractive has been shown to exhibit both electro-optic effects simultaneously.

4.3 Bright Soliton Formation

For bright solitons, $\rho = \frac{I_\infty}{I_d} = 0$. Hence, (4.7) becomes,

$$\begin{aligned} iU_\xi + \frac{1}{2}U_{ss} - \beta_2\frac{\left\{1 + \left(|U|^2\right)\exp\left[-\left(1 + |U|^2\right)\tau\right]\right\}^2}{\left(1 + |U|^2\right)^2}U \\ - \beta_1\frac{\left\{1 + \left(|U|^2\right)\exp\left[-\left(1 + |U|^2\right)\tau\right]\right\}}{\left(1 + |U|^2\right)}U = 0 \end{aligned} \tag{4.8}$$

The envelope U is expressed as, $U = r^{1/2}y(s)\exp(i\nu\xi)$ where ν is the nonlinear shift of the propagation constant, r is the intensity ratio, $r = \frac{I_{\max}}{I_d} = \frac{I(0)}{I_d}$ i.e., ratio of the maximum intensity to the dark irradiance; and $y(s)$ is a real normalized bounded function which satisfies $0 \leq y(s) \leq 1$. The boundary conditions for a bright soliton are, $y(0) = 1$, $y(s \to \pm\infty) = 0$, $\dot{y}(0) = 0$, $\dot{y}(\pm\infty) = 0$. Substitution U in (4.8) gives,

$$\frac{d^2y}{ds^2} = 2\nu y + 2\beta_2 \frac{\left\{1 + ry^2 \exp\left[-\left(1 + ry^2\right)\tau\right]\right\}^2 y}{\left(1 + ry^2\right)^2} + 2\beta_1 \frac{\left\{1 + ry^2 \exp\left[-\left(1 + ry^2\right)\tau\right]\right\} y}{\left(1 + ry^2\right)} \tag{4.9}$$

Equation (4.9) is a second order differential equation so integrating it once reduces it to a first order differential equation. Along with the bright soliton boundary conditions, we can obtain the value of the nonlinear shift of the propagation constant,

$$\nu = \int_1^0 \left[2\beta_1 \frac{\left\{1 + ry^2 \exp\left[-\left(1 + ry^2\right)\tau\right]\right\} y}{\left(1 + ry^2\right)} + 2\beta_2 \frac{\left\{1 + ry^2 \exp\left[-\left(1 + ry^2\right)\tau\right]\right\}^2 y}{\left(1 + ry^2\right)^2} \right] dy \tag{4.10}$$

Solving (4.9) numerically using (4.10) gives the spatial intensity profile of the soliton at a particular scaled time τ for a given r. It can be seen from (4.9) that the temporal evolution of the solitons depends explicitly on r and magnitude of both electro-optic coefficients. A broad bifurcation can now be made for low and high r to study the temporal evolution in detail.

4.3.1 Regime of High Intensity Ratio

Consider the intensity ratio r to be relatively high of the order of 10^2–10^3. As an illustration, solving (4.9) taking parameters of Table 4.1, we can derive the normalized spatial profiles for the intensity of the soliton at different time intervals as shown in Fig. 4.1.

Next, we shall study the temporal variation of the soliton width by plotting log-linear graph between the soliton width and the scaled time in Figs. 4.2, 4.3, 4.4, 4.5. The main concept behind the obtained curves can be understood as follows: Initially at $\tau = 0.005$, a soliton with a large spatial width starts to form because of the low magnitude of space charge field. As the induced space charge field builds up with the passage of time, the photorefractive effect saturates rapidly. Hence, the soliton's width now becomes minimum. This minimum width soliton is denoted to be a quasi-steady state soliton. Quasi steady state solitons have a historical precedent in that they were the first signs of self trapping in photorefractives back in the 1990s.

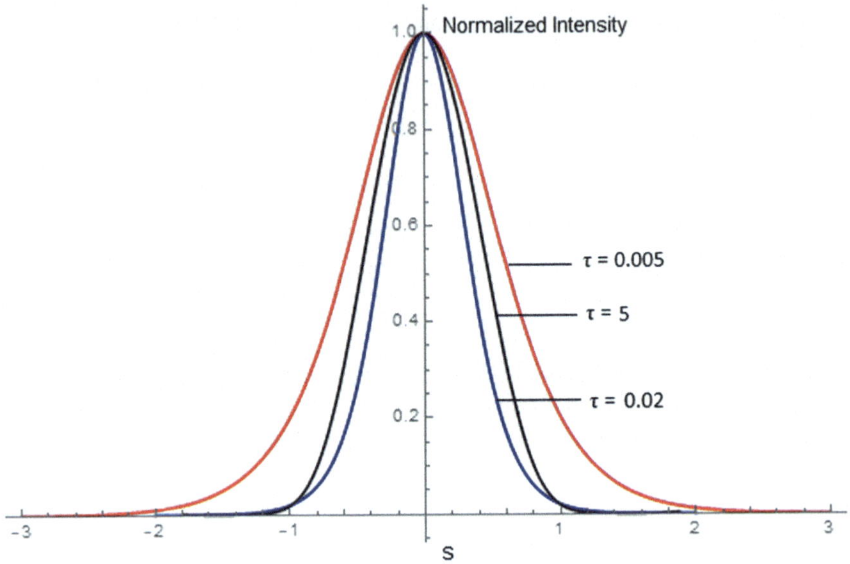

Fig. 4.1 Bright soliton formation at high intensity ratio r, $\tau = 0.005, 0.02, 5$ and $r = 25$. Reprinted from Chaos, Solitons and Fractals, 126, Aavishkar Katti, Temporal behaviour of bright solitons in photorefractive crystals having both the linear and quadratic electro-optic effect, 23–31, Copyright 2019, with permission from Elsevier

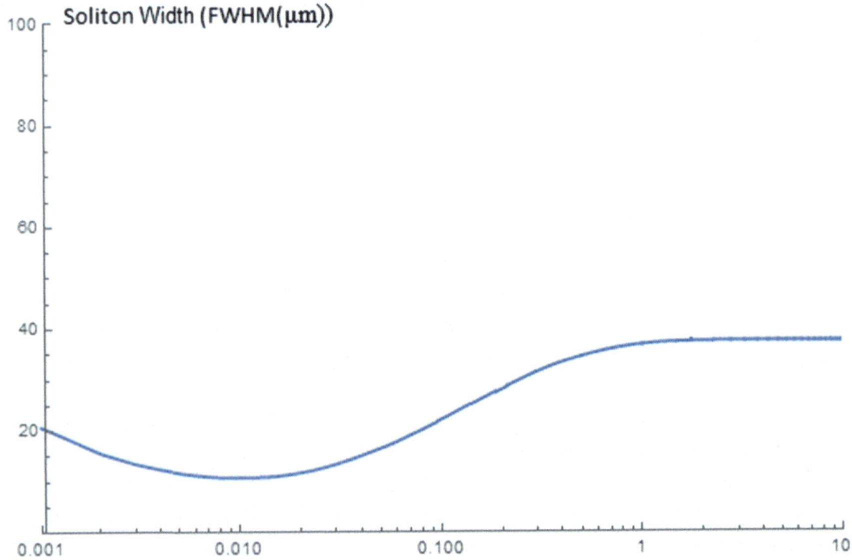

Fig. 4.2 Soliton width as a function of (scaled) time, $r = 200$. Reprinted from Chaos, Solitons and Fractals, 126, Aavishkar Katti, Temporal behaviour of bright solitons in photorefractive crystals having both the linear and quadratic electro-optic effect, 23–31, Copyright 2019, with permission from Elsevier

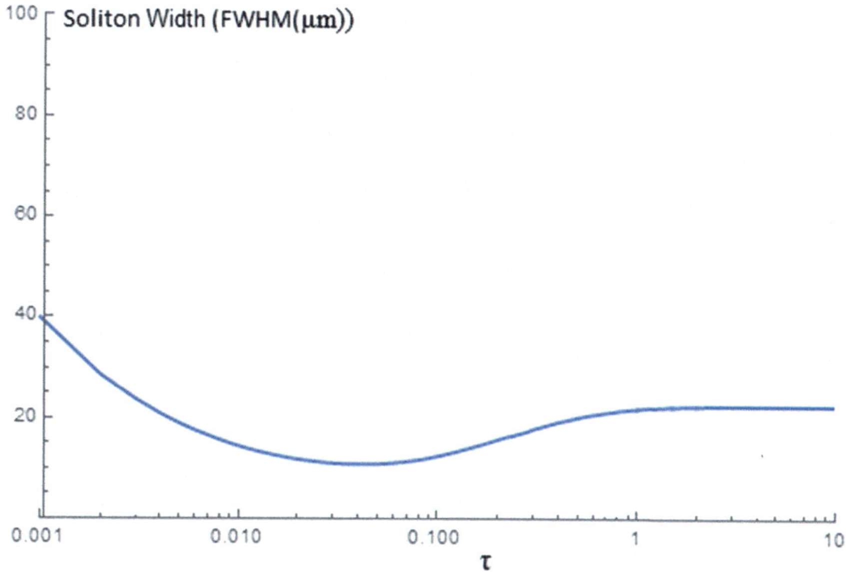

Fig. 4.3 Soliton width as a function of (scaled) time, $r = 50$. Reprinted from Chaos, Solitons and Fractals, 126, Aavishkar Katti, Temporal behaviour of bright solitons in photorefractive crystals having both the linear and quadratic electro-optic effect, 23–31, Copyright 2019, with permission from Elsevier

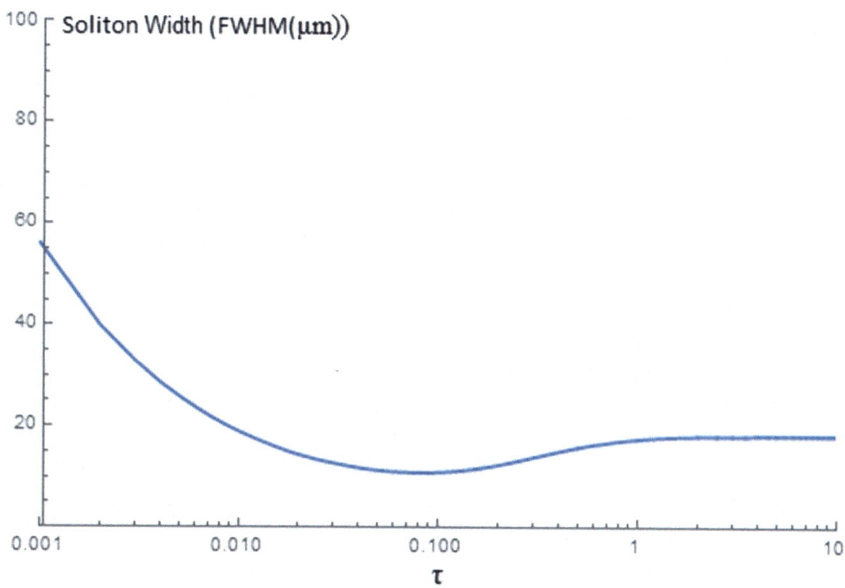

Fig. 4.4 Soliton width as a function of (scaled) time, $r = 25$. Reprinted from Chaos, Solitons and Fractals, 126, Aavishkar Katti, Temporal behaviour of bright solitons in photorefractive crystals having both the linear and quadratic electro-optic effect, 23–31, Copyright 2019, with permission from Elsevier

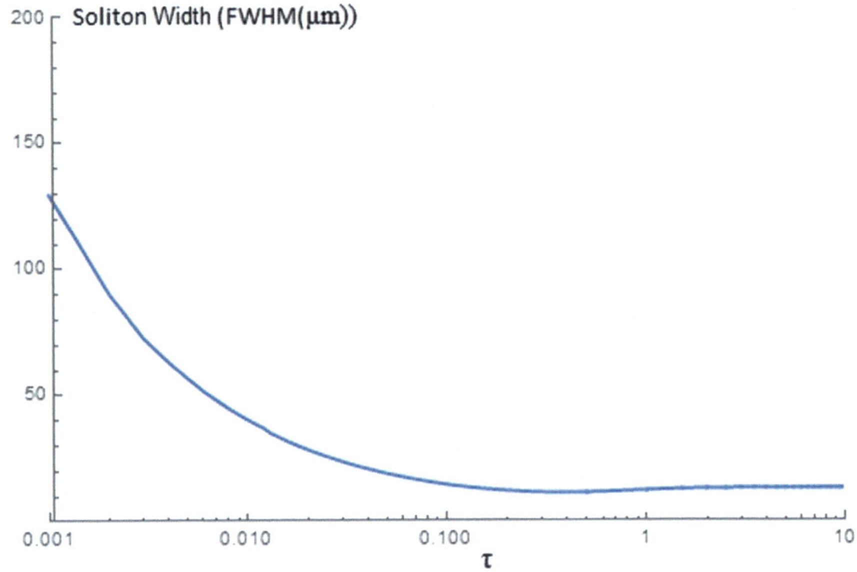

Fig. 4.5 Soliton width as a function of (scaled) time, $r = 5$. Reprinted from Chaos, Solitons and Fractals, 126, Aavishkar Katti, Temporal behaviour of bright solitons in photorefractive crystals having both the linear and quadratic electro-optic effect, 23–31, Copyright 2019, with permission from Elsevier

These quasi steady state solitons used to remain intact for a very short time interval and disperse after that. What is happening here is that the soliton width increases and reaches a perpetual value after $\tau = 1$. This implies that a steady state has been reached and the soliton width at this stage is called as the steady state soliton width. The change from the quasi steady state soliton to the steady state soliton happens when the screening of the external electric field commences and sets up an external screening field. Quasi steady state solitons exist in the narrow time interval between the time when the soliton width is minimum and the time at which the screening of the electric field starts.

There is an important inference to be made from Figs. 4.2, 4.3, 4.4 and 4.5. It is clear that the soliton width for the quasi steady state solitons approaches the soliton width for the steady state solitons with a decrease in the intensity ratio to approximately $r = 5$.

Also, it can be further inferred that with an increase in the intensity ratio r, the initial soliton width (say, at $\tau = 0.001$) decreases, while the steady state soliton width reduces with a reduction in the intensity ratio.

4.3.2 Regime of Low Intensity Ratio

We have seen earlier that the soliton width in case of quasi-steady state solitons comes closer and closer to the soliton width at steady state when we keep on decreasing the intensity ratio r till near about 5. So it is logical to now study how the soliton width changes with time for the intensity ratio $r < 5$. These plots of soliton width versus the scaled time are shown in Figs. 4.7, 4.8 and 4.9 while Fig. 4.6 shows the spatial soliton profiles for the intensity at different scaled time intervals for $r = 1$.

It is notable that quasi steady state solitons do not exist in the regime of low intensity ratio. The minimum value of the soliton width occurs in the steady state itself. The reason behind this is that the photorefractive effect does not saturate for low intensity ratios. For the current investigation, we consider $r < 5$ to be low and $r > 5$ to be high. The exact value of r at which we can bifurcate the low and high intensity regime will be slightly different for different photorefractive crystals because of the value of the linear or quadratic electro-optic coefficients and the interplay, if any, between the two electro-optic effects.

Also, the initial soliton width increases with a decrease in the intensity ratio r while the soliton width at steady state increases with a diminishing intensity ratio r.

Figure 4.10 shows the time to form for the quasi steady state solitons as a function of the intensity ratio. Since the slope of the line is found as ≈ 1, the time to form

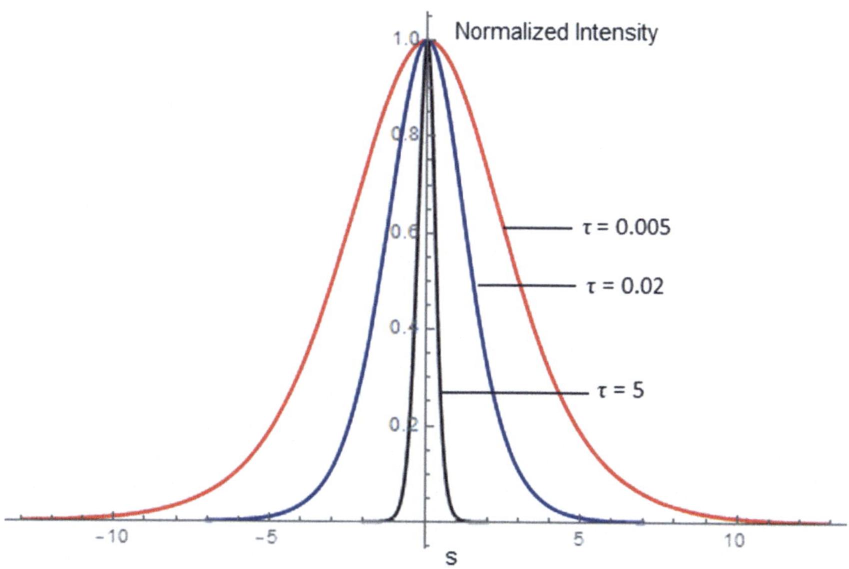

Fig. 4.6 Bright soliton formation in the regime of low intensity ratio r, $\tau = 0.005, 0.02, 5$ and $r = 1$. Reprinted from Chaos, Solitons and Fractals, 126, Aavishkar Katti, Temporal behaviour of bright solitons in photorefractive crystals having both the linear and quadratic electro-optic effect, 23–31, Copyright 2019, with permission from Elsevier

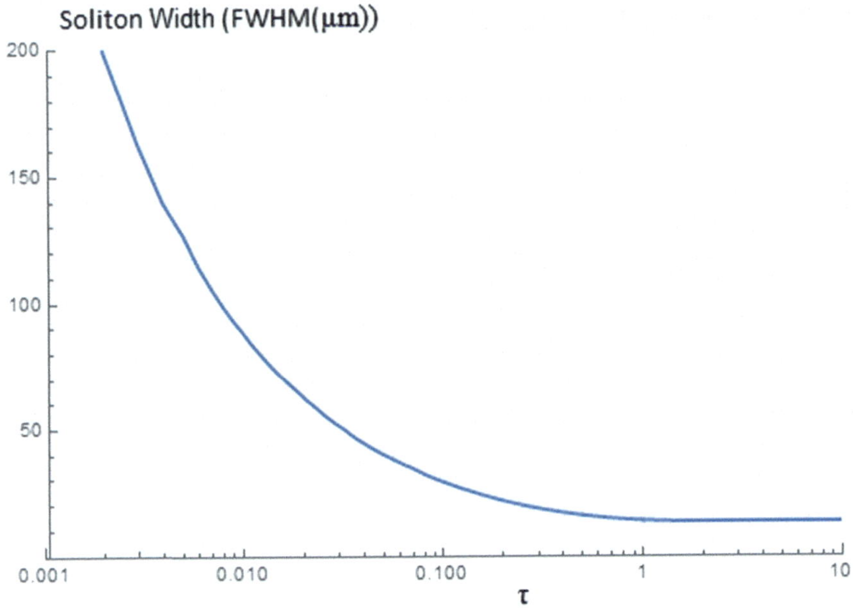

Fig. 4.7 Soliton width as a function of (scaled) time for $r = 1$. Reprinted from Chaos, Solitons and Fractals, 126, Aavishkar Katti, Temporal behaviour of bright solitons in photorefractive crystals having both the linear and quadratic electro-optic effect, 23–31, Copyright 2019, with permission from Elsevier

quasi steady state solitons τ_o is proportional to the inverse intensity ratio $1/r$, i.e., $r \sim (\tau_0)^{-1}$.

4.3.3 Influence and Interplay of Electro-optic Effects

In the previous section, the time evolution of the photorefractive solitons was investigated using parameters of PMN-0.33PT crystal which has the linear and quadratic electro-optic effect simultaneously. We shall now investigate how the magnitude of the electro-optic coefficients affect the formation characteristics. For such a study, consider two general cases. The first case is when the linear electro-optic effect much greater than the quadratic electro-optic effect, i.e., $\beta_1 \gg \beta_2$. The second case is when the quadratic electro-optic effect is much greater than the linear electro-optic effect, i.e., $\beta_1 \ll \beta_2$. Figure 4.11 shows the plot of the soliton width versus scaled time for both cases, $\beta_1 \gg \beta_2$ and $\beta_1 \ll \beta_2$ in the high intensity ratio regime.The soliton width at the initial time and the soliton width at steady state are lesser when the quadratic electro-optic effect dominates while the soliton width at quasi steady state is very nearly equal.

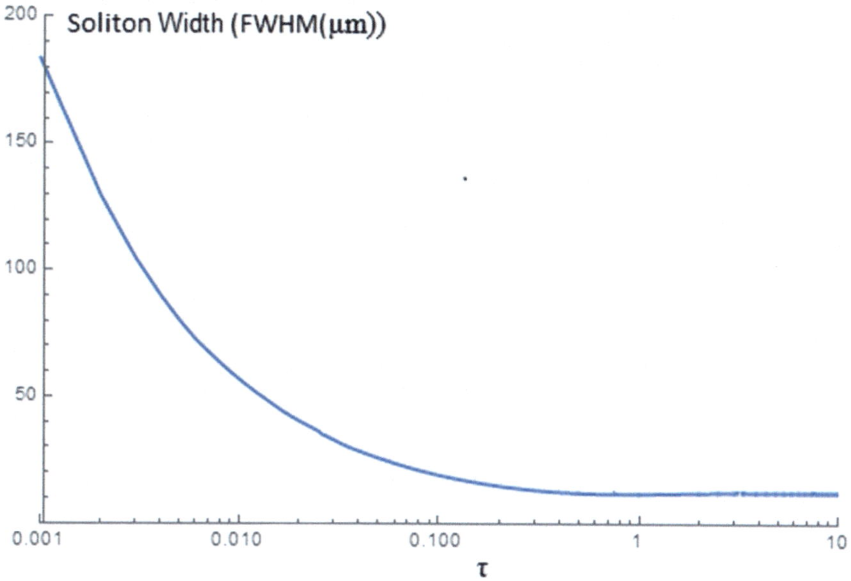

Fig. 4.8 Soliton width as a function of (scaled) time for $r = 2.5$. Reprinted from Chaos, Solitons and Fractals, 126, Aavishkar Katti, Temporal behaviour of bright solitons in photorefractive crystals having both the linear and quadratic electro-optic effect, 23–31, Copyright 2019, with permission from Elsevier

Figure 4.12 shows the plot of the soliton width versus scaled time for both cases, $\beta_1 \gg \beta_2$ and $\beta_1 \ll \beta_2$ in the low intensity regime. In this case, we can see that the soliton width at the initial time is much lesser when the quadratic electro-optic effect dominates. As discussed before, quasi steady state solitons cannot form if intensity ratio remains low.

4.3.4 Time Evolution of Screening Solitons

The theoretical foundation carries forward but with $\beta_2 = 0$ because now the photorefractive crystal is non-centrosymmetric and hence the quadratic electro-optic coefficient is zero. Hence, (4.8) becomes,

$$iU_\xi + \frac{1}{2}U_{ss} - \beta_1 \frac{\{1 + (|U|^2)\exp[-(1 + |U|^2)\tau]\}}{(1 + |U|^2)} U = 0 \qquad (4.11)$$

The ansatz for the field is again, $U = r^{1/2} y(s) \exp(i\nu\xi)$ where the symbols have their usual meaning as mentioned before. Substitution of U in (4.11) gives,

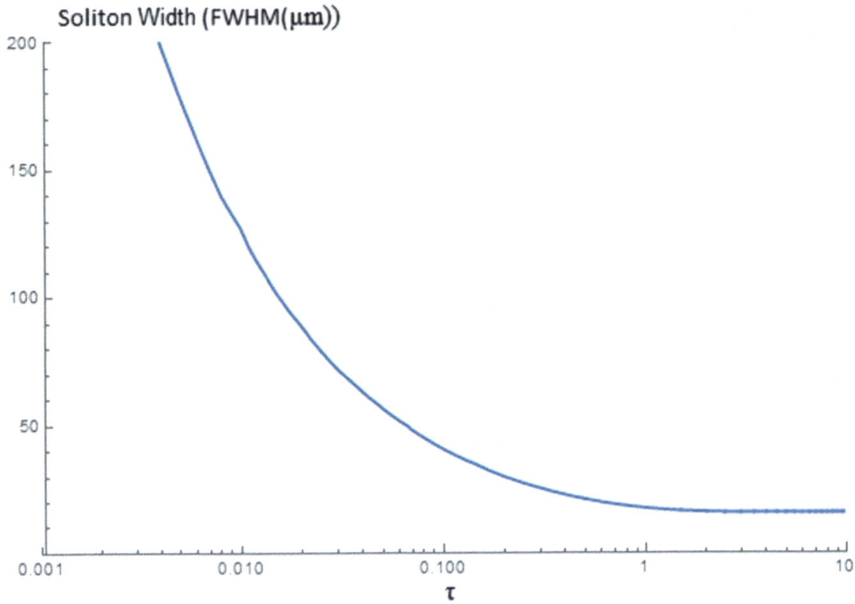

Fig. 4.9 Soliton width as a function of (scaled) time for $r = 0.5$. Reprinted from Chaos, Solitons and Fractals, 126, Aavishkar Katti, Temporal behaviour of bright solitons in photorefractive crystals having both the linear and quadratic electro-optic effect, 23–31, Copyright 2019, with permission from Elsevier

$$\frac{d^2y}{ds^2} = 2\nu y + 2\beta_1 \frac{\{1 + ry^2 \exp[-(1 + ry^2)\tau]\}y}{(1 + ry^2)} \tag{4.12}$$

Integrating (4.12) once, we obtain a first order differential equation. Using the bright soliton boundary conditions, we can obtain the nonlinear shift of the propagation constant,

$$\upsilon = \int_1^0 \left[2\beta_1 \frac{\{1 + ry^2 \exp[-(1 + ry^2)\tau]\}y}{(1 + ry^2)} \right] dy \tag{4.13}$$

Solving (4.12) numerically using (4.13) gives the spatial intensity profile of the soliton at a particular scaled time τ for a given r. Again, we can infer that the temporal characteristics of the bright screening solitons depend strongly on whether we are in the high or low intensity ratio regime. An analogous study to that done in Sects. 4.3.1, 4.3.2, 4.3.3 and 4.3.4 can be performed to firstly study the temporal characteristics at various intensity ratios and secondly, analyze the value of r below which quasi steady state solitons do not form.

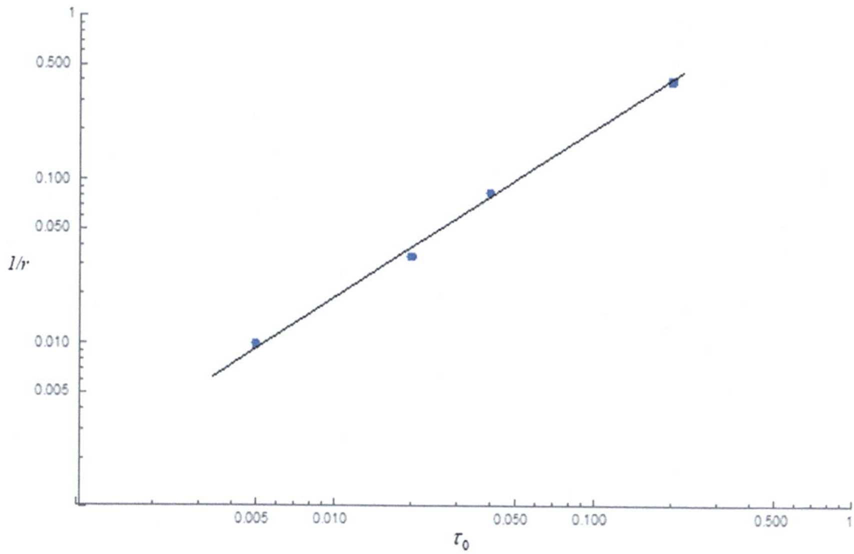

Fig. 4.10 Time to form the quasi steady state solitons versus the inverse intensity ratio (solid line = best fit, point = calculated). Reprinted from Chaos, Solitons and Fractals, 126, Aavishkar Katti, Temporal behaviour of bright solitons in photorefractive crystals having both the linear and quadratic electro-optic effect, 23–31, Copyright 2019, with permission from Elsevier

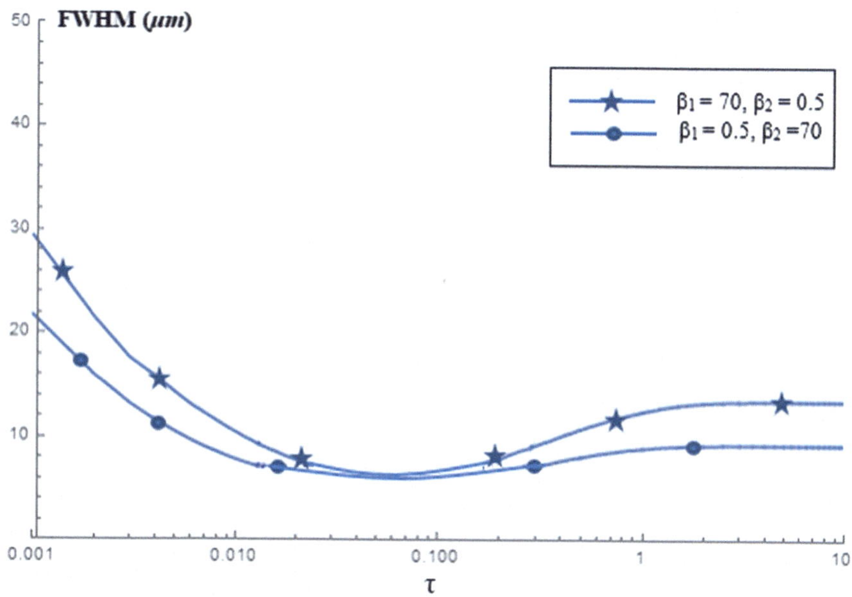

Fig. 4.11 Soliton width as a function of (scaled) time when $\beta_1 \gg \beta_2$ and $\beta_1 \ll \beta_2$ for $r = 25$ (high intensity ratio). Reprinted from Chaos, Solitons and Fractals, 126, Aavishkar Katti, Temporal behaviour of bright solitons in photorefractive crystals having both the linear and quadratic electro-optic effect, 23–31, Copyright 2019, with permission from Elsevier

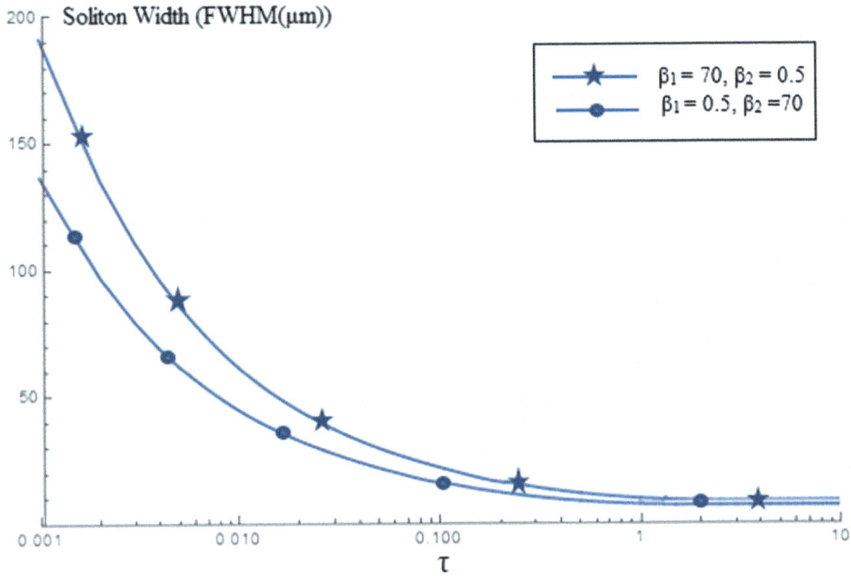

Fig. 4.12 Soliton width as a function of (scaled) time when $\beta_1 \gg \beta_2$ and $\beta_1 \ll \beta_2$ for $r = 0.5$ (low intensity ratio). Reprinted from Chaos, Solitons and Fractals, 126, Aavishkar Katti, Temporal behaviour of bright solitons in photorefractive crystals having both the linear and quadratic electro-optic effect, 23–31, Copyright 2019, with permission from Elsevier

4.3.5 Time Evolution of Centrosymmetric Solitons

The theoretical foundation elucidated in Eqs. (4.1)–(4.10) again carries forward but with $\beta_1 = 0$ because now the photorefractive crystal is centrosymmetric and hence the linear electro-optic coefficient is zero. Equation (4.8) for the case of bright solitons now becomes,

$$iU_\xi + \frac{1}{2}U_{ss} - \beta_2\frac{\{1 + (|U|^2)\exp[-(1 + |U|^2)\tau]\}^2}{(1 + |U|^2)^2}U = 0 \tag{4.14}$$

Substitution of $U = r^{1/2}y(s)\exp(i\nu\xi)$ in (4.14) leads to,

$$\frac{\mathrm{d}^2y}{\mathrm{d}s^2} = 2\nu y + 2\beta_2\frac{\{1 + ry^2\exp[-(1 + ry^2)\tau]\}^2 y}{(1 + ry^2)^2} \tag{4.15}$$

Integrating (4.15) once and using the bright soliton boundary conditions, we obtain,

$$v = \int_{1}^{0} \left[2\beta_2 \frac{\{1 + ry^2 \exp[-(1 + ry^2)\tau]\}^2 y}{(1 + ry^2)^2} \mathrm{d}y \right] \tag{4.16}$$

Solving (4.15) numerically using (4.16) gives the spatial intensity profile of the soliton at a particular scaled time τ for a given r. Again, we can infer that the temporal characteristics of the bright screening solitons depend strongly on whether the value of r is high or low. An analogous study to that done in Sects. 4.3.1, 4.3.2, 4.3.3 and 4.3.4 can be performed to firstly study the temporal characteristics at various intensity ratios and secondly, analyze the value of r below which quasi steady state solitons do not form.

4.4 Theoretical Formulation: Photovoltaic Photorefractive Crystals

Consider an optical beam propagating along the z-axis in a photorefractive crystal. The crystal is kept so that its c-axis coincides with the x axis and the diffraction is assumed to be in the same direction. The external electric field is applied along the x-axis and the light beam is linearly polarized along the same direction. In the slowly varying envelope approximation, the incident beam's electric field is, $E = \hat{x}A(x, z)\exp(ikz)$ where $k = k_0 n_e = (2\pi/\lambda_0)n_e$. λ_0 is the wavelength in free space and n_e is the unchanged refractive index. Hence, the dynamical evolution equation becomes,

$$\left(i\frac{\partial}{\partial z} + \frac{1}{2k}\frac{\partial^2}{\partial x^2} - \frac{k_0 n_e^3 r_{eff} E}{2} \right) A(x, z) = 0 \tag{4.17a}$$

where E is the induced space charge field. r_{eff} is the linear electro-optic coefficient. The amplitude term can now be expressed,

$$A(x, z) = u(x) I_d^{1/2} exp(i\Gamma z) \tag{4.17b}$$

The model of Kukhtarev which we detailed in the previous section will again be used to derive the time dependent space charge field but now the charge transport equations contain contribution from the bulk photovoltaic field as follows [13],

$$\frac{\partial}{\partial t} N_D^+ = (s_i I + \beta)(N_D - N_D^+) - \gamma n N_D^+ \tag{4.18a}$$

$$\frac{\partial}{\partial x}(\epsilon_0 \epsilon_r E) = \rho_i \tag{4.18b}$$

$$\frac{\partial J}{\partial x}+\frac{\partial \rho_i}{\partial t}=0 \tag{4.18c}$$

$$\rho_i=e\left(N_D^+-n-N_A\right) \tag{4.18d}$$

$$\hat{J}=e\mu nE_s+k_BT\mu\frac{\mathrm{d}n}{\mathrm{d}x}+\kappa s_i\left(N_D-N_D^+\right)I \tag{4.18e}$$

where the beam intensity is $I=|A|^2$, β is the generation rate of thermally generated or dark carriers, κ is the photovoltaic constant, the recombination rate of carriers is denoted by γ, s_{i} is the photoionization cross section, while the mobility of the electron is denoted by μ and e denotes the electron's charge. ρ_i is the density of the total charge while k_{B} represents the Boltzmann's constant. N_{A} and N_{D} are the concentrations of acceptors and donors respectively. n is the electron density, N_D^+ is the ionized donor density, $\hat{J}$ represents the electric current density and T is the temperature.

In principle, (4.2a)–(4.2e) can be solved simultaneously to obtain a time dependent space charge field. In practice, this is not so easy so a few approximations must be made to simplify the charge transport equations. Usually, the electron density is very small as compared to the density of ionized donors or acceptors if we consider moderate incident intensity and hence $N_D^+\approx N_A$. The carrier recombination time is negligible with respect to the dielectric response time which gives us $\partial N_D^+/\partial t=0$ [1]. Hence, (4.18d) becomes, $\rho\approx e(N_D^+-N_A)$. Substituting into the Gauss Law (4.18b), gives us, $\epsilon_0\epsilon_r\partial E/\partial x=e(N_D^+-N_A)$. If $x\to\pm\infty$, $E_s(x\to\pm\infty,z)=E_0$, then $\partial E_s/\partial x=0$ and $N_D^+=N_A$. (4.18a) elucidates the charge generation and recombination process and states that the characteristic recombination time for carriers is $1/\gamma n$. This is the response time for the buildup of free electron charge carriers. (4.18c) and (4.18e) are the continuity and current equations respectively. We can immediately infer that the dielectric response time of the buildup of ion charges is $\epsilon_r\epsilon_0/e\mu n$. Characteristically in photorefractive crystals, the recombination time is much smaller as compared to the dielectric response time, i.e., ${}^{1/\gamma n}/_{\epsilon_r\epsilon_0/e\mu n}\ll 1$ and hence only (4.18a) can be genuinely said to be a steady state equation. Under the above conditions,

$$n=\frac{s(N_D-N_A)(I+I_d)}{\gamma N_A} \tag{4.19}$$

where $I_d=\frac{\beta}{s_i}$. Putting (4.19) and $N_D^+\approx N_A$ into (4.18e),

$$\hat{J}=\frac{e\mu s_i(N_D-N_A)}{\gamma N_A}\left[(I+I_d)E_s+\frac{k_BT}{e}\frac{\partial I}{\partial x}+E_pI\right] \tag{4.20}$$

where, $E_P={}^{\kappa\gamma N_A}/_{e\mu}$ is known as the photovoltaic field constant. Substituting (4.20) and (4.18d) into (4.18b) gives,

$$T_d I_d \frac{\partial^2 E_s}{\partial x \partial t} + \frac{\partial[(I + I_d)E_s]}{\partial x} + \frac{k_B T}{e}\frac{\partial^2 I}{\partial x^2} + E_P \frac{\partial I}{\partial x} = 0 \tag{4.21}$$

where we have used, $T_d = (\in_0 \in_r / e\mu)[\gamma N_A / \beta(N_D - N_A)]$. Integrating (4.21), we heve,

$$T_d I_d \frac{\partial E_s}{\partial t} + (I + I_d)E_s + \frac{k_B T}{e}\frac{\partial I}{\partial x} + E_P I = C_1 \tag{4.22}$$

C_1 is a constant of integration. At steady state and in regions of constant illumination, $x \to \pm\infty$, $E_s(x \to \pm\infty, z) = E_0$, $I(x \to \pm\infty, z) = I_\infty$ and hence, the constant of integration can be found out from (4.22), $C_1 = (I_\infty + I_d)E_0 + E_p I_\infty$. Using the initial conditions that $E_s = 0$ and a slow variation of the intensity with respect to time, (4.22) can be solved,

$$E_s = \left(E_0 \frac{I_\infty + I_d}{I + I_d} - E_p \frac{I - I_\infty}{I + I_d} - \frac{k_B T}{e}\frac{1}{I + I_d}\frac{\partial I}{\partial x}\right)\left[1 - \exp\left(-\frac{I + I_d}{T_d I_d}t\right)\right] \tag{4.23}$$

E_0 can also be found out by using the potential condition,

$$V = -\int_{-l/2}^{l/2} E_s \mathrm{d}x = RS\hat{J} + \varepsilon \tag{4.24}$$

where V is the potential between the electrodes on the photorefractive crystal and are separated by a distance l. S is the surface area of the electrodes, ε represents the EMF of the source while R is the load resistance. Substituting (4.23) in (4.24), we get,

$$E_0 = -\left[\varepsilon/\chi + E_p \sigma/\chi - (k_B T/e)(\eta/\chi)\right] \tag{4.25}$$

where,

$$\chi = \int_{-l/2}^{l/2} \left\{\frac{(I_\infty + I_d)}{(I + I_d)}\right\}\left\{1 - \exp\left[-\frac{(I + I_d)}{(T_d I_d)}t\right]\right\}\mathrm{d}x$$

$$\sigma = \int_{-l/2}^{l/2} \left\{\frac{(I_\infty - I)}{(I + I_d)}\right\}\left\{1 - \exp\left[-\frac{(I + I_d)}{(T_d I_d)}t\right]\right\}\mathrm{d}x$$

$$\eta = \int_{-l/2}^{l/2} \left\{\frac{(\partial I/\partial x)}{(I + I_d)}\right\}\left\{1 - \exp\left[-\frac{(I + I_d)}{(T_d I_d)}t\right]\right\}\mathrm{d}x$$

Substituting (4.24) into (4.21), we have,

$$J = J_{0T} - \frac{1}{I_d(\varepsilon/\chi + E_p)} \left\{\left(E_p I + \frac{k_B T}{e}\frac{\partial I}{\partial x}\right)\left[1 - \exp\left(-\frac{I + I_d}{T_d I_d}t\right)\right] - \frac{k_B T}{e}\frac{\partial I}{\partial x} - E_p I\right\} \quad (4.26)$$

where,

$$J_{0T} = -\frac{1}{I_d(\varepsilon/\chi + E_p)} \left\{\frac{\varepsilon(I_\infty + I_d)}{\chi} + E_p\left[\frac{\sigma(I_\infty + I_d)}{\chi} - I_\infty\right] - \frac{k_B T \eta(I_\infty + I_d)}{e\chi}\right\} \quad (4.27)$$

Also, $J = \widehat{J}/[e\mu s(N_D - N_A)/\gamma N_A] I_d(\varepsilon/\chi + E_p)$ which implies that the electric current density has been transformed into dimensionless coordinates. The diffusion terms ($\frac{k_B T}{e}$ terms) can be neglected for a strong bias. Hence, the time dependent space charge field can be expressed as,

$$\widehat{E}_s = \frac{J_0 I_d - \alpha I}{I + I_d}\left[1 - \exp\left(-\frac{I + I_d}{T_d I_d}t\right)\right] \quad (4.28)$$

where, $J_0 = -\{1/[I_d(\varepsilon/\chi + E_p)]\}\{\varepsilon(I_\infty + I_d)/\chi + E_p[\sigma(I_\infty + I_d)/\chi - I_\infty]\}$, $\alpha = E_p/(\varepsilon/\chi + E_p)$, $\widehat{E}_s = E_s/(\varepsilon/\chi + E_p)$.

Following the approach of [8, 21], we shall now evaluate the value of J_0 for bright solitons. We shall assume that J_0 tends to approximately be equal to the steady state current density. We can infer this clearly from (4.26) taking the exponential term to be negligible. At steady state, the space charge field can be obtained by considering the exponential term to vanish,

$$\widehat{E}_s = (J_0 I_d - \alpha I)/(I + I_d) \quad (4.29)$$

Adopting the dimensionless co-ordinates,

$\xi = x/d, d = (\pm 2kb)^{-1/2}, b = (k/n_e)[(1/2)n_e^3 r_{eff}(\in/\chi + E_p)]$, where d is the scale length and the strength and sign of the nonlinearity can be inferred from b, and substituting in (4.24), we get,

$$J_0 = \beta C - \beta \int_{-l/2d}^{l/2d} \frac{J_0 - \alpha u^2}{1 + u^2} \mathrm{d}\xi \quad (4.30)$$

where, $\beta = \gamma N_A d/[RSI_d e\mu s(N_D - N_A)]$ and $C = \varepsilon/d(\varepsilon/\chi + E_p)$. Consider a bright soliton which has a width Δx and let u_0^2 maximum intensity. Approximating its shape by a square shaped function, and in the limit $l \gg \Delta x$, (4.17) becomes,

$$J_0 \cong \frac{\beta C(1+u_0^2) + (\beta \Delta x/d)\alpha u_0^2}{(1+u_0^2)[1+(\beta l/d)]} \tag{4.31}$$

Now, there are two limiting cases which are important to understand here. If the external resistance is very large or tends to infinity, i.e., $R \to \infty$, then $\beta \to 0$ and $J_0 = 0$. If the external resistance is vanishingly small or nonexistent, i.e., $R \to 0$, $J_0 \to [\varepsilon(1+u_0^2)/(\varepsilon/\chi + E_p) + \alpha \Delta x u_0^2]/[(1+u_0^2)/l]$ and if $u_0^2 << 1$, or $u_0^2 >> 1$ (4.21) reduces to,

$$J_0 = \frac{\varepsilon}{l(\varepsilon/\chi + E_p)} + \frac{\alpha \Delta x}{l} \tag{4.32}$$

We can now obtain the dynamical evolution equation by using (4.17) and (4.28),

$$\frac{d^2u}{d\xi^2} = \pm\left\{\frac{\Gamma}{b} + \frac{J_0 - \alpha u^2}{1+u^2}\left[1 - \exp\left(-\frac{1+u^2}{T_d}t\right)\right]\right\}u \tag{4.33}$$

Integrating (4.33) once, we get,

$$\left(\frac{du}{d\xi}\right)^2 = \pm\left\{\begin{array}{l} \left(\frac{\Gamma}{b} - \alpha\right)\left(u^2 - u_0^2\right) + (J_0+\alpha)\ln\left(\frac{1+u^2}{1+u_0^2}\right) \\ -(J_0+\alpha)\left[Ei\left(-\frac{1+u^2}{T_d}t\right) - Ei\left(-\frac{1+u_0^2}{T_d}t\right)\right] \\ -\frac{\alpha T_d}{t}\left[\exp\left(-\frac{1+u^2}{T_d}t\right) - \exp\left(-\frac{1+u_0^2}{T_d}t\right)\right] \end{array}\right\}u \tag{4.34}$$

For bright soliton solutions, the boundary conditions are,

$$u_\infty = u'(\infty) = u''(\infty) = 0, u'(0) = 0.$$

So, using the boundary condition at infinity,

$$\begin{aligned} \frac{\Gamma}{b} = \alpha + \frac{J_0+\alpha}{u_0^2}\left[\ln\left(\frac{1}{1+u_0^2}\right) - Ei\left(-\frac{1}{T_d}t\right) + Ei\left(-\frac{1+u_0^2}{T_d}t\right)\right] \\ - \frac{\alpha T_d}{u_0^2 t}\left[\exp\left(-\frac{1}{T_d}t\right) - \exp\left(-\frac{1+u_0^2}{T_d}t\right)\right] \end{aligned} \tag{4.35}$$

Substituting (4.35) into (4.34), we finally get,

$$\left(\frac{\mathrm{d}u}{\mathrm{d}\xi}\right)^2 = \left((J_0+\alpha)\left\{ \begin{array}{l} \ln(1+u^2) + Ei\left(-\frac{1}{T_d}t\right) - Ei\left(-\frac{1+u_0^2}{T_d}t\right) \\ -\frac{u^2}{u_0^2}\left[\ln(1+u_0^2) + Ei\left(-\frac{1}{T_d}t\right) - Ei\left(-\frac{1+u_0^2}{T_d}t\right)\right] \end{array} \right\} + \frac{\alpha T_d}{t}\left\{ \begin{array}{l} \exp\left(-\frac{1}{T_d}t\right) - \exp\left(-\frac{1+u^2}{T_d}t\right) \\ -\frac{u^2}{u_0^2}\left[\exp\left(-\frac{1}{T_d}t\right) - \exp\left(-\frac{1+u_0^2}{T_d}t\right)\right] \end{array} \right\} \right) \tag{4.36}$$

The bright soliton's temporal evolution can be found out by numerically integrating (4.36). As an example, we shall consider $l = 1$ cm, R = 0 and typical $LiNbO_3$ parameters [8] with $E_p = 40$ kV/cm. Figure 4.13 shows the variation of the soliton FWHM with t/T_d for $u_0^2 = 10$ when ε = 30,000; 20,000; 10,000; −10,000; −20,000 V for $J_0 = 0.43, 0.33, 0.20, -0.33, -0.99$ respectively. Figure 4.14 shows how the soliton width changes with t/T_d for ε = 20,000 V with $J_0 = 0.33$ when $u_0^2 = 0.1, 1, 10$ respectively. From Figs. 4.13 and 4.14, it is evident that the time to form for the steady-state bright solitons reduces with an increase in u_0^2 when electric current density J_0 is constant. Also, for a constant u_0^2, $J_0 > 0$ is needed to form for steady-state bright solitons as the minimum formation time for steady state solitons does not exist for $J_0 < 0$.

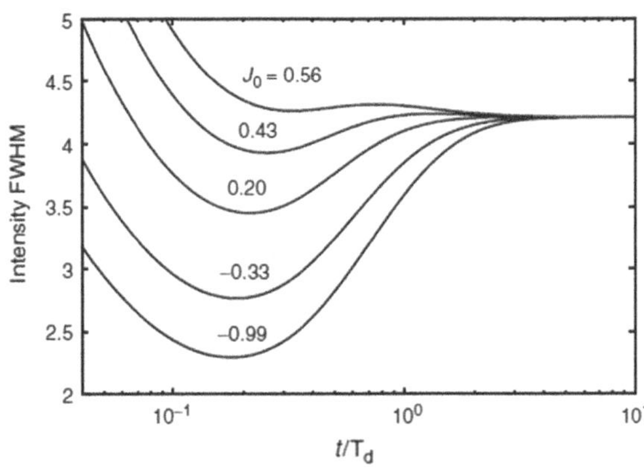

Fig. 4.13 Soliton width versus t/T_d for $u_0^2 = 10$ when ε = 30,000; 20,000; 10,000; −10,000; −20,000 V for $J_0 = 0.43, 0.33, 0.20, -0.33, -0.99$. Temporal development of spatial solitons in biased photorefractive-photovoltaic materials, Keqing Lu, Wei Zhao et al., Journal of Modern Optics, Vol. 55, 10 June 2008, Taylor and Francis, reprinted by permission of the publisher (Taylor &Francis Ltd, https://www.tandfonline.com)

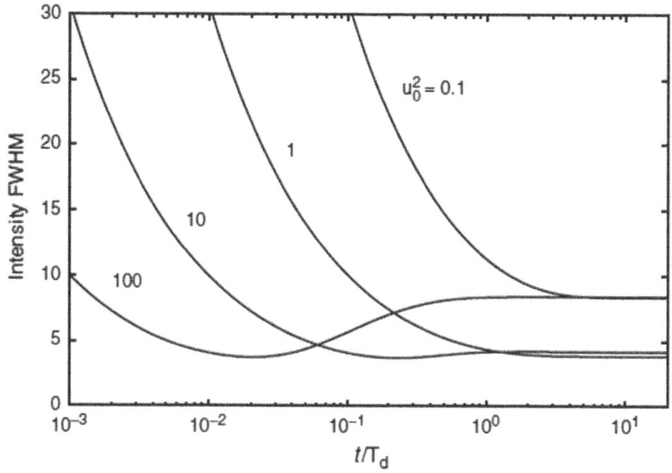

Fig. 4.14 Soliton Width as a function of t/T_d for $\varepsilon = 20{,}000$ V with $J_0 = 0.33$ when $u_0^2 = 0.1, 1, 10$. Temporal development of spatial solitons in biased photorefractive-photovoltaic materials, Keqing Lu, Wei Zhao et al., Journal of Modern Optics, Vol. 55, 10 June 2008, Taylor and Francis, reprinted by permission of the publisher (Taylor &Francis Ltd, https://www.tandfonline.com)

4.5 Further Reading

In 4.4, when the external bias field is absent, $\varepsilon = 0$ and the system reduces to that of photovoltaic solitons and (4.36) is modified accordingly by changing the value of J_0 to be that at $\varepsilon = 0$ [as found in (4.28)]. If the photovoltaic effect is negligible, $E_{\mathrm{p}} = 0$, and the system becomes that of screening solitons for $R = 0$. Substituting $E_{\mathrm{p}} = 0$ and modifying J_0 for $E_{\mathrm{p}} = 0$ in (4.36), we can get the soliton envelope variation with time. Reference [8] treats these aspects relating to temporal evolution of screening photovoltaic solitons in detail, including the evolution of both bright and dark solitons.

In addition, Refs. [5–7, 9, 12] should be referred for an in depth understanding of temporal dynamics of solitons in different types of photorefractive crystals.

References

1. N. Fressengeas, J. Maufoy, G. Kugel, Temporal behavior of bidimensional photorefractive bright spatial solitons. Phys. Rev. E—Stat. Phys. Plasmas, Fluids Relat. Interdiscip. Top. **54**(6), 6866–6875 (1996). https://doi.org/10.1103/PhysRevE.54.6866
2. J. Maufoy, N. Fressengeas, D. Wolfersberger, G. Kugel, Simulation of the temporal behavior of soliton propagation in photorefractive media. Phys. Rev. E **59**(5), 6116 (1999)
3. N. Fressengeas, D. Wolfersberger, J. Maufoy, G. Kugel, Experimental study of the self-focusing process temporal behavior in photorefractive Bi12TiO20. J. Appl. Phys. **85**(4), 2062–2067 (1999)

4. N. Fressengeas, D. Wolfersberger, J. Maufoy, G. Kugel, Build up mechanisms of (1+ 1)-dimensional photorefractive bright spatial quasi-steady-state and screening solitons. Opt. Commun. **145**(1–6), 393–400 (1998)
5. M. Chauvet, Temporal analysis of open-circuit dark photovoltaic spatial solitons. JOSA B **20**(12), 2515–2522 (2003)
6. Z. Lei, L. Ke-Qing, Z. Mei-Zhi, L. Xue-Min, Z. Yan-Peng, Temporal development of open-circuit bright photovoltaic solitons. Chin. Phys. B. **17**(7), 2539 (2008)
7. K. Lu et al., Temporal behavior of dark spatial solitons in closed-circuit photovoltaic media. Opt. Commun. **281**(10), 2913–2917 (2008)
8. K. Lu et al., Temporal development of spatial solitons in biased photorefractive-photovoltaic materials. J. Mod. Opt. **55**(10), 1571–1585 (2008)
9. K. Zhan, C. Hou, S. Pu, Temporal behavior of spatial solitons in centrosymmetric photorefractive crystals. Opt. Laser Technol. **43**(7), 1274–1278 (2011)
10. A. Ziółkowski, Temporal analysis of solitons in photorefractive semiconductors. J. Opt. **14**(3), 035202 (2012). https://doi.org/10.1088/2040-8978/14/3/035202
11. D. Wolfersberger, N. Khelfaoui, C. Dan, N. Fressengeas, H. Leblond, Fast photorefractive self-focusing in InP: Fe semiconductor at infrared wavelengths. Appl. Phys. Lett. **92**(2), 021106 (2008)
12. A. Katti, Temporal behaviour of bright solitons in photorefractive crystals having both the linear and quadratic electro-optic effect. Chaos, Solitons Fractals **126**, 23–31 (2019)
13. N.V. Kukhtarev, V.B. Markov, S.G. Odulov, M.S. Soskin, V.L. Vinetskii, Holographic storage in electrooptic crystals. I. steady state. Ferroelectrics **22**(1), 949–960 (1978). https://doi.org/10.1080/00150197908239450
14. D.N. Christodoulides, M.I. Carvalho, Bright, dark, and gray spatial soliton states in photorefractive media. J. Opt. Soc. Am. B **12**(9), 1628 (1995). https://doi.org/10.1364/JOSAB.12.001628
15. M. Wesner, C. Herden, R. Pankrath, D. Kip, P. Moretti, Temporal development of photorefractive solitons up to telecommunication wavelengths in strontium-barium niobate waveguides. Phys. Rev. E **64**(3), 036613 (2001)
16. A. Katti, R.A. Yadav, A. Prasad, Bright optical spatial solitons in photorefractive waveguides having both the linear and quadratic electro-optic effect. Wave Motion **77**, 64–76 (2018). https://doi.org/10.1016/J.WAVEMOTI.2017.10.002
17. X. Wan, D.Y. Wang, X. Zhao, H. Luo, H.L.W. Chan, C.L. Choy, Electro-optic characterization of tetragonal (1 − x) Pb (Mg1/3Nb2/3) O3–xPbTiO 3 single crystals by a modified Sénarmont setup. Solid State Commun. **134**(8), 547–551 (2005). https://doi.org/10.1016/j.ssc.2005.02.033
18. A. Katti, Incoherently coupled Gaussian soliton pairs in biased photorefractive crystal having both the linear and quadratic electro-optic effect. Appl. Phys. B **124**(9), 192 (2018)
19. R. Zhang, B. Jiang, W. Cao, Elastic, piezoelectric, and dielectric properties of multidomain 0.67Pb(Mg1/3Nb2/3)O3–0.33PbTiO3 single crystals. J. Appl. Phys. **90**(7), 3471–3475 (2001). https://doi.org/10.1063/1.1390494
20. Q. Jiang, Y. Su, H. Nie, Z. Ma, Y. Li, Separate spatial soliton pairs in a biased series photorefractive crystal circuit with both the linear and quadratic electro-optic effects. J. Mod. Opt. **64**(6), 609–615 (2016)
21. M. Segev, G.C. Valley, M.C. Bashaw, M. Taya, M.M. Fejer, Photovoltaic spatial solitons. J. Opt. Soc. Am. B **14**(7), 1772 (1997). https://doi.org/10.1364/JOSAB.14.001772

Chapter 5
Coupling of Photorefractive Solitons

5.1 Introduction

Coupling of optical solitons has been an attractive topic for research since around 1960s [1]. The pairing of optical spatiotemporal solitons has been studied previously in Kerr media by solving two coupled NLS equations. The methodology was then extended to optical spatial solitons. There has been previous research which theoretically predicts bright and/or dark spatial soliton pairs containing two different wavelengths can be observed in a self-focusing or defocusing medium with a Kerr type nonlinearity by cross phase modulation. The two superimposed soliton components should have requisitely scaled relative intensities [2]. As an experimental confirmation of the above, a bright–dark spatial soliton pair has been detected and observed in a self focusing nonlinear medium with the Kerr nonlinearity with the two constituent beams being of different colors [3]. In general, any two beams propagating collinearly in a nonlinear medium will interact through cross phase modulation. Hence one beam affects the other through cross phase modulation. In case of coupled soliton pair, if the two components have equivalent amplitude, cross phase modulation is almost equivalent with the self-phase modulation of each beam, and both effects play a vital role in the formation of a soliton pair. Therefore, one component of the soliton pair cannot exist independently of the other. This is in direct contrast to the case where a weak beam is guided by a induced waveguide due to a strong soliton beam. The cross phase modulation is much smaller than the self phase modulation and the weaker beam has relatively less effect on the stronger soliton beam in such a case. If the weak beam is switched off, the stronger beam is not affected [4].

If we consider one dimensional spatial solitons, then there are three scenarios in which they can interact with each other in a plane. The two solitons can either overlap, propagate parallel to each other some distance apart, or cross each other. If two mutually coherent self trapped beams are propagating parallel to each other in a photorefractive crystal but not overlapping, then an attraction or repulsion is observed between the two solitons, depending on their initial phase difference. In this chapter, we shall be investigating the interaction between two incoherent or

A. Katti and R. A. Yadav, *Optical Spatial Solitons in Photorefractive Materials*, Progress in Optical Science and Photonics 14, https://doi.org/10.1007/978-981-16-2550-3_5

coherent overlapping, collinearly propagating photorefractive solitons leading to coupled spatial soliton pairs. We can consider two mutually incoherent or mutually coherent light beams possessing the same frequency and polarization. These two light beams are responsible for an effective refractive index modulation being produced in the photorefractive crystal. Each soliton beam is self trapped due to the refractive index waveguide induced by a combination of intensities of both beams. Hence, the coupled pair will collapse if any one light beam is switched off. In case there are more than two incident soliton beams, this theory can be extended to study incoherently coupled multicomponent solitons. The first theoretical studies on incoherently coupled spatial solitons were conducted by Christodoulides et al. [40] and the experimental studies followed [5, 6]. Consequently, many investigations have been performed for incoherently coupled soliton pairs and coupled multicomponent solitons in numerous realizations.

The experimental apparatus for possible realization of incoherent or coherent coupled soliton pairs and multicomponent solitons consists firstly, of a CW laser in the range 400–600 nm. Then a beam splitter is used split the laser beam into two soliton-like beams. If one wishes to have multicomponent coupled solitons, a series of beam splitters can be used to form 2 N soliton like beams. A glass slide is inserted between them. If one wishes to observe an incoherently coupled soliton pair, these beams are made mutually incoherent at the crystal face by having their optical path difference to be very much greater than the coherence length of the laser. For a coherently coupled soliton pair, the optical path difference between the two beams should be constant and within the coherence length of the laser. The output beam is detected by a CCD camera.

Since the collinearly propagating soliton beams have the same frequency and polarization, the way to distinguish one component in the output from the others is by blocking the other components by means of a mechanical shutter and sampling the desired soliton beam in a time interval less than the dielectric response time of the crystal. What this does is it allows one to view the desired component of the soliton, as a rapid change in intensity prevents any effect on the effective refractive index induced due to all the beams [6, 7].

In this chapter, a theoretical foundation will be presented for both incoherent and coherent coupled soliton pairs in photorefractive media. Finally, taking an approximate solution for the coupled soliton pair reveals an entirely different parameter space for unique Gaussian soliton pair which we then investigate in detail. The theory behind incoherently and coherently coupled soliton pairs can be easily generalized to study multicomponent incoherently or coherently coupled solitons which we briefly comment upon at the end of the chapter.

5.2 Theoretical Foundation: Photovoltaic Photorefractive Crystals

Consider two incoherent light beams which are propagating along the z axis and which can diffract along the x axis only. The two light light beams possess linear polarization along the x axis while the external electric field is also applied in the same direction. The total electric field for the two beams can now be written as, $\overrightarrow{E} = \overrightarrow{E}_1 + \overrightarrow{E}_2$.The total electric field satisfies the following Helmholtz equation,

$$\nabla^2 \vec{E} + \left(k_0 n_e'\right) \vec{E} = 0 \tag{5.1}$$

where the symbols have their usual meaning as discussed previously. The individual electric fields are expressed in terms of slowly varying envelopes, $E_1 = i\phi(x, z)\exp(ikz)$ and $E_2 = i\psi(x, z)\exp(ikz)$ where $k = k_0 n_e$. Substituting these envelopes into the Helmholtz Eq. (5.1) along with the expression for the perturbed extraordinary refractive index $\Delta n = n'_e - n_e = -\frac{1}{2} n_e^3 r_{eff} E_{sc}$, we get,

$$i\phi_z + \frac{1}{2k}\phi_{xx} - \frac{k_0\left(n_e^3 r_{eff} E_{sc}\right)}{2}\phi = 0 \tag{5.2}$$

$$i\psi_z + \frac{1}{2k}\psi_{xx} - \frac{k_0\left(n_e^3 r_{eff} E_{sc}\right)}{2}\psi = 0 \tag{5.3}$$

where, $\phi_z = \partial\phi/\partial z$, etc.

Now the next step for solving (5.2) and (5.3) would be to calculate the space charge field. The space charge field has already been obtained for photovoltaic photorefractive crystals in previous chapters (see, for example, 3.1),

$$\begin{aligned}
E_{sc} = &\left(E_0 \frac{1}{1+|U|^2} - E_p \frac{|U|^2}{1+|U|^2} + E_p \frac{|U|^2}{1+|U|^2} f L_D \frac{\partial}{\partial x}\frac{E_{sc}}{E_t}\right) \\
&\left(1 + L_D \frac{\partial}{\partial x}\frac{E_{sc}}{E_t}\right)\left(1 - f L_D \frac{\partial}{\partial x}\frac{E_{sc}}{E_t}\right)^{-1} \\
&- \frac{K_b T}{e}\left\{\frac{\partial}{\partial x}\ln(1+|U|^2) - \left[\left(1 + L_D \frac{\partial}{\partial x}\frac{E_{sc}}{E_t}\right)^{-1}\right.\right. \\
&\left.\left. + f\left(1 - f L_D \frac{\partial}{\partial x}\frac{E_{sc}}{E_t}\right)^{-1}\right] L_D \frac{\partial^2}{\partial x^2}\frac{E_{sc}}{E_t}\right\}
\end{aligned} \tag{5.4}$$

where, $L_D = (\epsilon_0 \epsilon_r k_B T / e^2 N_A)^{1/2}$ is the Debye length or the diffusion length, T is the temperature, $E_t = \frac{e N_A L_D}{\epsilon_0 \epsilon_r}$, $f = N_A/(N_D - N_A)$ and $E_p = \kappa\gamma N_A/e\mu$ which is the photovoltaic field constant. μ is the electron mobility and other symbols have their usual meaning as defined before. If the intensity variation is relatively smooth, the diffusion effect can be neglected relative to the photovoltaic effects in typical photorefractive materials. Hence, the term $L_D \frac{\partial E_{sc}}{\partial x}$ has a value much less than unity [8, 9] and (5.4) reduces to,

$$E_{sc} = E_0 \frac{I_\infty + I_d}{I + I_d} + E_p \frac{I_\infty - I}{I + I_d} \tag{5.5}$$

E_0 can be obtained from the potential condition,

$$\epsilon = -\int_{-l/2}^{l/2} E dx \tag{5.6}$$

ϵ is the external voltage applied to the crystal between the electrodes which are separated by a distance l.

Using (5.5) and (5.6), we get,

$$E_{sc} = -(\epsilon\eta + E_p \widehat{\sigma}\eta) \frac{I_\infty + I_d}{I + I_d} + E_p \frac{I_\infty - I}{I + I_d} \tag{5.7}$$

where,

$$\eta = 1 / \int_{-l/2}^{l/2} [(I_\infty + I_d)/(I + I_d)] dx$$
$$\hat{\sigma} = -\int_{-l/2}^{l/2} \frac{(I_\infty - I)}{(I + I_d)} dx$$

If the two beams are coupled incoherently, i.e., there is no relation between the phases of the two beams at any instant, then the total intensity of the two beams can be found out by adding the two Poynting fluxes,

$I = \frac{n_e}{2\eta_0}\left(|\phi|^2 + |\psi|^2\right)$ with $\eta_0 = (\mu_0/\epsilon_0)^{1/2}$. Employing the usual dimensionless coordinates where the intensity is scaled with respect to the dark irradiance, $\phi = (2\eta_0 I_d/n_e)^{1/2} U$, $\psi = (2\eta_0 I_d/n_e)^{1/2} V$, $\xi = z/(kx_0^2)$, $s = x/x_0$, the normalized envelopes U and V now satisfy,

$$iU_\xi + \frac{1}{2} U_{ss} + (\alpha + \beta)(\rho + 1) \frac{U}{1 + |U|^2 + |V|^2} - \delta \frac{[\rho - (|U|^2 + |V|^2)]U}{1 + |U|^2 + |V|^2} = 0 \tag{5.8}$$

$$iV_\xi + \frac{1}{2} V_{ss} + (\alpha + \beta)(\rho + 1) \frac{V}{1 + |U|^2 + |V|^2} - \delta \frac{[\rho - (|U|^2 + |V|^2)]V}{1 + |U|^2 + |V|^2} = 0 \tag{5.9}$$

where, $\beta = (k_0^2 x_0^2)(n_e^4 r_{33} \eta/2)\epsilon$, $\alpha = (k_0^2 x_0^2)(n_e^4 r_{33} \widehat{\sigma} \eta/2) E_p$, $\delta = (k_0^2 x_0^2)(n_e^4 r_{33}/2) E_p$.

In case the two beams do have a phase relationship at all instants in time, they constitute a coherently coupled soliton pair. The total intensity now becomes, $I = \frac{n_e}{2\eta_0}(|\phi + \psi|^2)$and the dynamical evolution equations are,

$$iU_\xi + \frac{1}{2}U_{ss} + (\alpha + \beta)(\rho + 1)\frac{U}{1 + |U + V|^2} - \delta\frac{[\rho - (|U + V|^2)]U}{1 + |U + V|^2} = 0 \quad (5.10)$$

$$iV_\xi + \frac{1}{2}V_{ss} + (\alpha + \beta)(\rho + 1)\frac{V}{1 + |U + V|^2} - \delta\frac{[\rho - (|U + V|^2)]V}{1 + |U + V|^2} = 0 \quad (5.11)$$

It is notable to observe here that the above theory works when the bulk photovoltaic effect is negligible, i.e., $E_p = 0$. In this case, Eq. (5.5) becomes,

$$E_{sc} = E_0\frac{I_\infty + I_d}{I + I_d} \quad (5.12)$$

Substituting (5.12) in (5.2) and (5.3), the system reduces to the case of studying coupled screening soliton pairs in nonphotovoltaic photorefractive solitons. Now, we know the following relations for the current for a photovoltaic photorefractive crystal at steady state, (see Note 1, Chap. 3),

$$J = e\mu n E_{sc} + k_B T\mu\frac{\partial n}{\partial x} + k_p s_i(N_D - N_D^+)I \quad (5.13)$$

$$\frac{\partial J}{\partial x} = 0 \quad (5.14)$$

$$n = \frac{s_i(N_D - N_A)(I + I_d)}{\gamma_R N_A}\left(1 + L_D\frac{\partial}{\partial x}\frac{E_{sc}}{E_t}\right)^{-1} \quad (5.15)$$

(5.13) and (5.14) imply that $J = J_\infty$ and hence,

$$J = J_\infty = e\mu n E_0 + k_p s_i(N_D - N_D^+)I_\infty \quad (5.16)$$

Substituting the value of E_0 from the potential condition as found in (5.7),

$$J = -e\mu n(\epsilon\eta + E_p\widehat{\sigma}\eta) + k_p s_i(N_D - N_D^+)I_\infty \quad (5.17)$$

If there is no external bias, $\epsilon = 0$, and using $E_p = \kappa\gamma N_A/e\mu$, $N_D^+ \sim N_A$ with $N_A/N_D \ll 1$ in typical photorefractive materials, we get from (5.13)–(5.15),

$$\frac{J}{sI_d N_D k_p} = \widehat{J} = \frac{I_\infty - \widehat{\sigma}\eta(I_\infty + I_d)}{I_d} \quad (5.18)$$

Substituting in (5.7) with $\epsilon = 0$ with $I = |U|^2 I_d = u^2 I_d$, we obtain,

$$\frac{E_{sc}}{E_p} = \frac{\widehat{J} - u^2}{1 + u^2} \tag{5.19}$$

Hence, in (5.19), we have obtained the space charge field for the case of photovoltaic photorefractive crystals without the external bias field. Now, the photovoltaic crystal can be open circuit or closed circuit. For an open circuit, $\widehat{J} = 0$ and (5.19) becomes,

$$E_{sc} = -E_p \frac{I/I_d}{1 + I/I_d} \tag{5.20}$$

Substituting (5.19) in (5.2) and (5.3), we obtain a system for studying the coupled soliton pairs in closed circuit photovoltaic crystals without an external bias field. Similarly, substituting (5.20) in (5.2) and (5.3), we obtain a system for studying the coupled soliton pairs in open circuit photovoltaic crystals without an external bias field.

5.3 Theoretical Foundation: Non-photovoltaic Photorefractive Crystals

Now, in non photovoltaic photorefractive materials, the change in extraordinary refractive index is [8, 10–12],

$$\Delta n = n'_e - n_e = -\frac{1}{2} a n_e^3 r_{eff} E_{sc} - \frac{1}{2} b n_e^3 g_{eff} \varepsilon_0^2 (\varepsilon_r - 1)^2 E_{sc}^2 \tag{5.21}$$

where E_{sc} is the space charge field induced in the photorefractive crystal. r_{eff} represents the linear electro-optic coefficient, g_{eff} represents the quadratic electro-optic coefficient while ε_0 and ε_r are the permittivity in vacuum and the dielectric constant respectively.

a and b are nonzero depending upon whether the photorefractive crystal exhibits the linear ($a = 1, b = 0$) or quadratic ($a = 0, b = 1$) both ($a = 1, b = 1$) electro-optic effects.

Simplifying (5.21) to the first order, we get,

$$(n_e^/)^2 = n_e^2 - n_e^4 r_{eff} E_{sc} - n_e^4 g_{eff} \varepsilon_0^2 (\varepsilon_r - 1)^2 E_{sc}^2 \tag{5.22}$$

The total electric field is expressed as $\vec{E} = \vec{E_1} + \vec{E_2}$. E_1 and E_2 are the individual components while the total field satisfies the Helmholtz equation. The slowly varying envelopes are, $E_1 = i\phi(x, z)\exp(ikz)$ and $E_2 = i\psi(x, z)\exp(ikz)$ where $k = k_0 n_e$. Substituting these envelopes into the Helmholtz Eq. (5.1) and using (5.22), we get,

$$\left(i\frac{\partial}{\partial z} + \frac{1}{2k}\frac{\partial^2}{\partial x^2} + \frac{k}{n_e}\Delta n\right)\phi(x, z) = 0 \tag{5.23}$$

$$\left(i\frac{\partial}{\partial z} + \frac{1}{2k}\frac{\partial^2}{\partial x^2} + \frac{k}{n_e}\Delta n\right)\psi(x, z) = 0 \tag{5.24}$$

with

$$\Delta n = -\frac{1}{2}an_e^3 r_{eff} E_{sc} - \frac{1}{2}bn_e^3 g_{eff}\varepsilon_0^2(\varepsilon_r - 1)^2 E_{sc}^2 \tag{5.25}$$

For a relatively broad beam and strong bias, the static space charge field has been obtained previously by neglecting the effects of diffusion,

$$E_{sc} = E_0\frac{I_\infty + I_d}{I + I_d} \tag{5.26}$$

where the symbols have their usual meanings. For finding E_0, refer to the circuit shown in Fig. 5.1 from where the following potential condition is obtained,

$$\varepsilon = -\int_{-l/2}^{l/2} E_{sc}dx + RJS \tag{5.27}$$

where ε is the EMF of the source, R is an external load, S is surface area of the electrodes, and l is the width of the crystal between the electrodes [11]. Substitution of (5.27) into (5.26) gives,

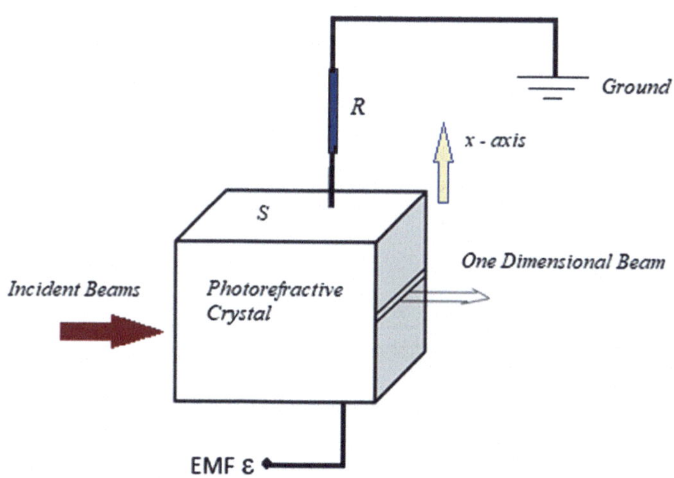

Fig. 5.1 Electrical circuit for observing coupled solitons in a biased photorefractive crystal. R is the external resistance

$$E_{sc} = -(\varepsilon\alpha - RJS\alpha)\frac{I_\infty + I_d}{I + I_d} \tag{5.28}$$

where,

$$\alpha = \frac{1}{\int_{-l/2}^{l/2}(I_\infty + I_d)(I + I_d)^{-1}dx} \tag{5.29}$$

The total intensity for the two mutually incoherent beams is gotten by adding the two Poynting fluxes, $I = \frac{n_e}{2\eta_0}\left(|A_1|^2 + |A_2|^2\right)$ with $\eta_0 = (\mu_0/\varepsilon_0)^{1/2}$.

Substituting E_{sc} and Δn in (5.2) and in terms of dimensionless variables specified (see 5.2), one gets the following equation,

$$iU_\xi + \frac{1}{2}U_{ss} - \frac{\beta_1(1+\rho)U}{1+|U|^2+|V|^2} - \frac{\beta_2(1+\rho)^2U}{(1+|U|^2+|V|^2)^2} \tag{5.30}$$

$$iV_\xi + \frac{1}{2}V_{ss} - \frac{\beta_1(1+\rho)V}{1+|U|^2+|V|^2} - \frac{\beta_2(1+\rho)^2V}{(1+|U|^2+|V|^2)^2} = 0 \tag{5.31}$$

and $\beta_1 = (k_0x_0)^2n_e^4r_{eff}\frac{(RJS\alpha-\varepsilon\alpha)^2}{2}$, $\beta_2 = (k_0x_0)^2n_e^4g_{eff}\varepsilon_0^2(\varepsilon_r-1)^2\frac{(RJS\alpha-\varepsilon\alpha)^2}{2}$

5.4 Coupled Spatial Solitons

Considering first the case of a photovoltaic photorefractive crystal and a bright incoherently coupled soliton pair, let us define the normalized envelopes [9],

$$U = r^{1/2}y(s)\cos\theta\exp(i\mu\xi) \tag{5.32}$$

$$V = r^{1/2}y(s)\sin\theta\exp(i\mu\xi) \tag{5.33}$$

where, θ is an arbitrary projection, μ represents the nonlinear shift of the propagation constant, $y(s)$ is a real bounded function such that $0 \le y(s) \le 1$. Substituting (5.32) and (5.33) in (5.10) and (5.11),

$$\frac{d^2y}{ds^2} - 2\mu y + 2\delta\frac{ry^3}{1+ry^2} + 2(\alpha+\beta)\frac{y}{1+ry^2} = 0 \tag{5.34}$$

With the boundary conditions for bright solitons to be $y(0) = 1$, $y'(0) = 0$, $y(s \to \pm\infty) = 0$

Integrating (5.34) once and using the boundary conditions elucidated above,

Table 5.1 Typical parameters used for Lithium Niobate crystal in the calculation [13]

Parameter	Value	Parameter	Value
λ_0	0.5 μm	l	1 cm
x_0	40 μm	E_p	40 kV/cm
n_e	2.2	r	10
r_{33}	30×10^{-12} mV^{-1}	θ	π/6
ϵ	10,000 V		

$$\mu = -\frac{\delta - \beta - \alpha}{r}\ln(1 + r) + \delta \tag{5.35}$$

$$\left(\frac{dy}{ds}\right)^2 = 2\frac{(\delta - \beta - \alpha)}{r}[-y^2\ln(1 + r) + \ln(1 + ry^2)] \tag{5.36}$$

Further integration gives the soliton profile,

$$[2(\delta - \beta - \alpha)]^{1/2}s = \pm\int_y^1 \frac{r^{1/2}d\hat{y}}{[\ln(1 + r\hat{y}^2) - \hat{y}^2\ln(1 + r)]^{1/2}} \tag{5.37}$$

To illustrate a bright incoherently coupled soliton pair, we consider a typical lithium niobate crystal whose parameters are shown in Table 5.1 and the normalized spatial profile is shown in Fig. 5.2.

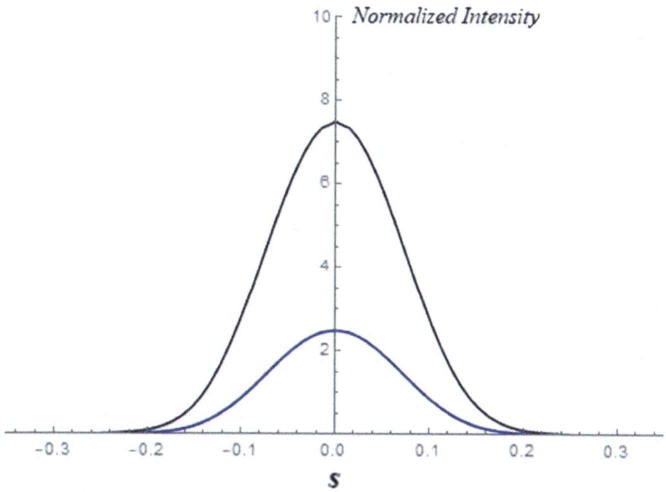

Fig. 5.2 Normalized intensity profile for bright–bright soliton pair predicted by (5.37) when $r = 10$, $\theta = \pi/6$, $\epsilon = 10{,}000$ V [9]

Similarly, now considering the case of a non-photovoltaic photorefractive crystal and a bright incoherently coupled soliton pair, let us define the normalized envelopes [14],

$$U = r^{1/2} y(s) \cos\theta \exp(i\mu\xi) \quad (5.38)$$

$$V = r^{1/2} y(s) \sin\theta \exp(i\mu\xi) \quad (5.39)$$

where $r = I(0)/I_d = I_{max}/I_d$, μ represents the nonlinear shift of the propagation constant, θ represents an arbitrary projection, *y(s)* is a real bounded function such that $0 \le y(s) \le 1$. Substituting (5.38) and (5.39) in (5.30) and (5.31), we get,

$$\left(\frac{dy}{ds}\right)^2 - 2\nu y - 2\beta_1 \frac{y}{1+ry^2} - 2\beta_2 \frac{y}{(1+ry^2)^2} = 0 \quad (5.40)$$

Integrating (5.31) twice, we can get the spatial profile as follows,

$$s = \pm \int_y^1 \frac{d\tilde{y}}{\left\{\frac{2\beta_1}{r}\left[\ln(1+r\tilde{y}^2) - \tilde{y}^2 \ln(1+r)\right] + \frac{2\beta_2}{1+r}\frac{r\tilde{y}^2(1-\tilde{y}^2)}{1+r\tilde{y}^2}\right\}^{1/2}} \quad (5.41)$$

$$\nu = -\frac{\beta_1}{r}\ln(1+r) - \frac{\beta_2}{1+r} \quad (5.42)$$

Let us consider a PMN-0.33 PT crystal which exhibits both the linear and quadratic electro-optic effect simultaneously. Parameters taken are mentioned in Table 5.2 and the normalized spatial profile is shown in Fig. 5.3.

Lastly, we shall consider the coherently coupled bright soliton pair in photorefractive media. Since the beams are now coupled together along with a mutual phase relationship between them, the normalized envelopes now become [15],

$$V = r_2^{1/2} y(s) \exp(i\mu\xi + \theta_2) \quad (5.43)$$

Table 5.2 Typical parameters used for PMN-0.33PT crystal in the calculation for incoherently coupled soliton pairs [12]

Parameter	Value	Parameter	Value
λ_0	632.8 nm	l	1 cm
x_0	40 μm	E_0	5×10^5 V/m
n_e	2.562	r	10
r_{33}	182×10^{-12} mV^{-1}	β_1	309.23
$g_{eff}\epsilon_0^2(\epsilon_r - 1)^2$	1.36×10^{-16} m^2/V^2	β_2	115.79

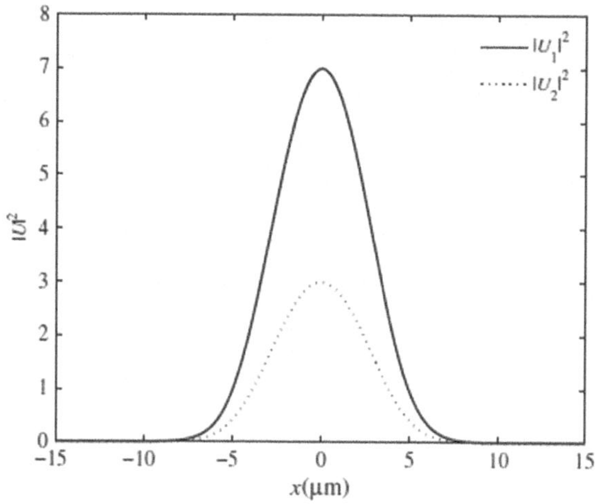

Fig. 5.3 Normalized intensity profiles of the coupled soliton pair predicted in (5.41) when $E_0 = 5 \times 10^5$ V/m and $r = 10$. (Reprinted From "Incoherently coupled spatial soliton pairs due to both the linear and quadratic electro-optic effects", Lili Hao, Qiang Wang, et al., Journal of Modern Optics, 62, 6 February 2015, reprinted by permission of the publisher (Taylor & Francis Ltd., https://www.tandfonline.com)

$$U = r_1^{1/2} y(s) \exp(i\mu\xi + \theta_1) \tag{5.44}$$

where $r_i = I_i(0)/I_d = (I_i)_{max}/I_d$; $i = 1, 2$, μ represents the nonlinear shift of the propagation constant, $\theta_l$and θ_2 represent the phases of the two beams respectively, $y(s)$ is a real bounded function such that $0 \leq y(s) \leq 1$. Substituting (5.43) and (5.44) in (5.30) and (5.31), we get,

$$\left(\frac{dy}{ds}\right)^2 - 2\nu y - 2\beta_1 \frac{y}{1 + ry^2} - 2\beta_2 \frac{y}{(1 + ry^2)^2} = 0 \tag{5.45}$$

Integrating (5.45) twice, we can get the spatial profile as follows,

$$s = \pm \int_y^1 \frac{d\tilde{y}}{\left\{\frac{2\beta_1}{r}\left[\ln(1 + r\tilde{y}^2) - \tilde{y}^2 \ln(1 + r)\right] + \frac{2\beta_2}{1+r}\frac{r\tilde{y}^2(1-\tilde{y}^2)}{1+r\tilde{y}^2}\right\}^{1/2}} \tag{5.46}$$

$$\nu = -\frac{\beta_1}{r}\ln(1 + r) - \frac{\beta_2}{1 + r} \tag{5.47}$$

with,

$r = r_1 + r_2 + 2\sqrt{r_1 r_2}\cos(\Delta\theta)$ and $\Delta\theta = \theta_2 - \theta_1$ being the phase difference between the two beams. We again consider the PMN-0.33PT crystal to illustrate the coherently coupled bright soliton pair. Table 5.3 encapsulates the parameters taken in this particular calculation and Fig. 5.4 shows the normalized spatial profiles of the coherently coupled soliton pair. Now, since there is mutual phase difference between the two beams, it is evident that the nonlinearity will be affected by the modification in mutual phase difference. Hence, in Fig. 5.5, we plot how the spatial width of the

Table 5.3 Typical parameters used for PMN-0.33PT crystal in the calculation for coherently coupled soliton pairs [12, 15]

Parameter	Value	Parameter	Value
λ_0	632.8 nm	l	1 cm
x_0	40 μm	E_0	3×10^5 V/m
n_e	2.562	r_1	5
r_{33}	182×10^{-12} $\mathrm{mV^{-1}}$		
$g_{eff}\epsilon_0^2(\epsilon_r - 1)^2$	1.36×10^{-16} $\mathrm{m^2/V^2}$		

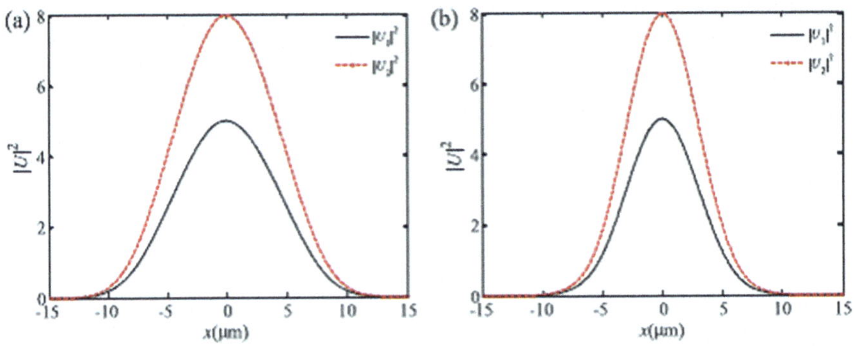

Fig. 5.4 Normalized intensity profiles for the coherently coupled soliton pair when **a** $\Delta\theta = \pi/4$ **b** $\Delta\theta = 3\pi/4$ and $E_0 = 3 \times 10^5$ V/m. (Reprinted from Optik---International Journal for Light and Electron Optics, 127, Lili Hao, Chunfeng Hou,Qiang Wang, Haiwei Mu, Coherently coupled spatial soliton pairs in biased photorefractive crystals with both the linear and quadratic electro-optic effects, 4339–4344, Copyright 2016, with permission from Elsevier)

soliton pair behaves for different values of $\Delta\theta$. Figure 5.6 shows the variation of the spatial width with the beam intensity ratio r_2/r_1.

5.5 Gaussian Soliton Pairs

Recently, a novel type of soliton pair has been investigated known as "Gaussian" soliton pairs or "quasi-soliton" pairs. The primary difference between these soliton pairs and the previously considered soliton pairs is the fact that these soliton pairs have a Gaussian profile. It is notable that photorefractive solitons are neither Gaussian nor hyperbolic secant and usually we have to perform numerical calculations to solve the paraxial Helmholtz equation for obtaining the spatial profile. However, in Refs. [16–21], it has been shown how the Gaussian ansatz can used to solve the paraxial Helmholtz equation and is a good approximation to the exact numerical solutions. Such Gaussian soliton pairs uncover a unique parameter space for existence of a new type of coupled soliton pairs which possess distinct characteristics relating

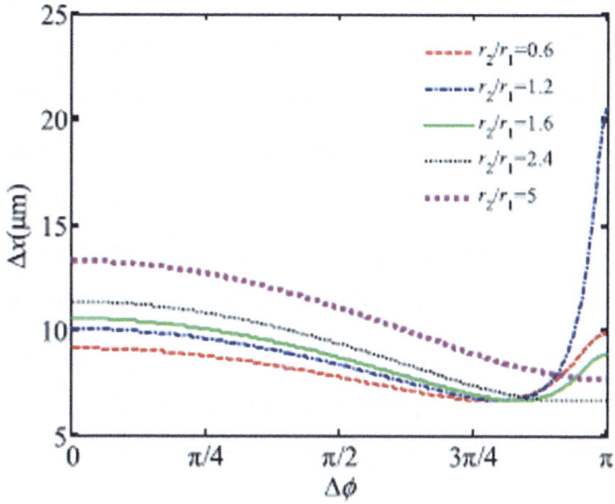

Fig. 5.5 Coherently coupled bright soliton pair width as a function of phase difference $\Delta\theta$ for various beam intensity ratios. $E_0 = 3 \times 10^5$ V/m. (Reprinted from Optik—International Journal for Light and Electron Optics, 127, Lili Hao, Chunfeng Hou,Qiang Wang,Haiwei Mu, Coherently coupled spatial soliton pairs in biased photorefractivecrystals with both the linear and quadratic electro-optic effects, 4339–4344, Copyright 2016, with permission from Elsevier) (Reprinted by permission from Springer Nature: Springer Applied Physics B: Lasers and Optics, Incoherently coupled Gaussian soliton pairs in biased photorefractive crystal having both the linear and quadratic electro-optic effect, Aavishkar Katti, Copyright 2018)

Fig. 5.6 The spatial width of the coherently coupled soliton pair as a function of the beam intensity ratio for different mutual phase differences. (Reprinted by permission from Springer Nature: Springer Applied Physics B: Lasers and Optics, Incoherently coupled Gaussian soliton pairs in biased photorefractive crystal having both the linear and quadratic electro-optic effect, Aavishkar Katti, Copyright 2018)

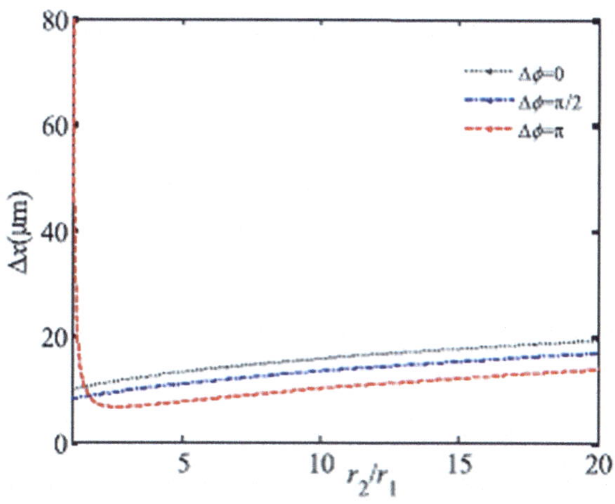

to dynamical evolution and stability. Since these pertain to approximate solutions, these solitons should be understood to be "quasi-solitons". It is notable that a full numerical simulation has indeed confirmed the validity of such an approximation [16]. To begin an investigation into these quasi soliton pairs, let us recapitulate the evolution equations first,

$$i\frac{\partial U}{\partial \xi}+\frac{1}{2}\frac{\partial^2 U}{\partial s^2}-\beta_1\frac{1+\rho}{1+|U|^2+|V|^2}U-\beta_2\left(\frac{1+\rho}{1+|U|^2+|V|^2}\right)^2U=0 \quad (5.48)$$

$$i\frac{\partial V}{\partial \xi}+\frac{1}{2}\frac{\partial^2 V}{\partial s^2}-\beta_1\frac{1+\rho}{1+|U|^2+|V|^2}V-\beta_2\left(\frac{1+\rho}{1+|U|^2+|V|^2}\right)^2V=0 \quad (5.49)$$

where
$\beta_1 = a\frac{(k_0x_0)^2 n_e^4 r_{eff}}{2}E_0$, $\beta_2 = b\frac{(k_0x_0)^2 n_e^4 g_{eff}\epsilon_0^2(\epsilon_r-1)^2}{2}E_0^2$ with $\rho = I_\infty/I_d$.

The value of a and b can be one or zero depending upon the nonlinearity exhibited by the photorefractive crystal. Equation (5.48) and (5.49) cannot be solved to obtain exact solution. Hence, one has to resort to numerical methods and there are many different types of approximate methods which work quite well for this. Segev's method [22], Akhmanov's paraxial method [23], Anderson's variational method [24] and Vlasov's moment method [25] are some of the approximate methods which can be used. The paraxial approximation shall be considered using a variational solution in order to obtain soliton states. Of course, for bright solitons, $\rho = 0$. Additionally, the slowly varying beam envelope is expressed in terms of a general plane wave ansatz,

$$U_1(\xi, s) = A_1(\xi, s)e^{-i\nu_1(s,\xi)} \quad (5.50)$$

$$U_2(\xi, s) = A_2(\xi, s)e^{-i\nu_2(s,\xi)} \quad (5.51)$$

$A_1(\xi, s)$ and $A_2(\xi, s)$ are real functions and $\nu(\xi, s)$ represents the phase. Using (5.50) and (5.51) in (5.34) and (5.35) gives,

$$\frac{\partial A_1}{\partial \xi}-\frac{\partial A_1}{\partial s}\frac{\partial \nu_1}{\partial s}-\frac{1}{2}A_1\frac{\partial^2 \nu_1}{\partial s^2}=0 \quad (5.52)$$

$$\frac{\partial A_2}{\partial \xi}-\frac{\partial A_2}{\partial s}\frac{\partial \nu_2}{\partial s}-\frac{1}{2}A_2\frac{\partial^2 A_2}{\partial s^2}=0 \quad (5.53)$$

$$A_1\frac{\partial \nu_1}{\partial \xi}+\frac{1}{2}\frac{\partial^2 A_1}{\partial s^2}-\frac{1}{2}A_1\left(\frac{\partial \nu_1}{\partial s}\right)^2-\beta_1\Phi_1(\xi,s)A_1-\beta_2\Phi_2(\xi,s)A_1=0 \quad (5.54)$$

$$A_2\frac{\partial \nu_2}{\partial \xi}+\frac{1}{2}\frac{\partial^2 A_2}{\partial s^2}-\frac{1}{2}A_2\left(\frac{\partial \nu_2}{\partial s}\right)^2-\beta_1\Phi_1(\xi,s)A_2-\beta_2\Phi_2(\xi,s)A_2=0 \quad (5.55)$$

where,

$$\Phi_1(\xi, s) = \frac{1}{1+|A_1|^2+|A_2|^2} \quad (5.56)$$

$$\Phi_2(\xi, s) = \left(\frac{1}{1+|A_1|^2+|A_2|^2}\right)^2 \quad (5.57)$$

$\Phi_1(\xi, s)$ and $\Phi_2(\xi, s)$ represent the non-linearity due to the linear and quadratic electro-optic effect. These terms represent counteraction of diffraction leading to formation of a stable soliton.

Now, we need to solve for obtaining bright spatial soliton solutions for (5.52)–(5.55). Since, these equations are non integrable, modified gaussian solutions will be used as approximate solutions [24, 26–31]. The quasi-soliton ansatz will be taken as,

$$A_1(\xi, s) = \frac{\sqrt{P_1}}{\sqrt{f_1(\xi)}}\exp\left[\frac{-s^2}{2r_1^2 f_1^2(\xi)}\right] \tag{5.58}$$

$$\nu_1(\xi, s) = \frac{s^2}{2}\eta_1(\xi) + \phi_1(\xi) \tag{5.59}$$

$$\eta_1(\xi) = -\frac{1}{f_1(\xi)}\frac{df_1(\xi)}{d\xi} \tag{5.60}$$

$$A_2(\xi, s) = \frac{\sqrt{P_2}}{\sqrt{f_2(\xi)}}\exp\left[\frac{-s^2}{2r_2^2 f_2^2(\xi)}\right] \tag{5.61}$$

$$\nu_2(\xi, s) = \frac{s^2}{2}\eta_2(\xi) + \phi_2(\xi) \tag{5.62}$$

$$\eta_2(\xi) = -\frac{1}{f_2(\xi)}\frac{df_2(\xi)}{d\xi} \tag{5.63}$$

where P_1 and P_2 signify the peak powers of these solitons, r_1, $f_1(\xi)$, $f_2(\xi)$, $\eta_1(\xi)$, $\eta_2(\xi)$are the variational parameters needed to define a solution in (5.58) and (5.61). The product $r_1 f_1(\xi)$, $r_2 f_2(\xi)$ indicates the spatial width of the respective solitons. Expanding Φ_1, Φ_2 by Taylor series expansion and approximating to the first order,

$$\Phi_1 \cong \frac{1}{(1 + (P_1/f_1) + (P_2/f_2))} + s^2\left(\frac{P_1}{r_1^2 f_1^3} + \frac{P_2}{r_2^2 f_2^3}\right)\frac{1}{(1 + (P_1/f_1) + (P_2/f_2))^2} \tag{5.64}$$

$$\Phi_2 \cong \frac{1}{(1 + (P_1/f_1) + (P_2/f_2))^2} + 2s^2\left(\frac{P_1}{r_1^2 f_1^3} + \frac{P_2}{r_2^2 f_2^3}\right)\frac{1}{(1 + (P_1/f_1) + (P_2/f_2))^3} \tag{5.65}$$

Using (5.58)–(5.65) in (5.54) and (5.55), we obtain a polynomial equation which has terms of various powers of s. Equating the coefficients of s^2 of the ensuing equation, we have,

$$\frac{d^2 f_1}{d\xi^2} = \frac{1}{r_1^4 f_1^3} - 2\beta_1\left(\frac{P_1}{r_1^2 f_1^2} + \frac{P_2 f_1}{r_2^2 f_2^3}\right)\frac{1}{\left(1 + \frac{P_1}{f_1} + \frac{P_2}{f_2}\right)^2}$$

$$-4\beta_2\left(\frac{P_1}{r_1^2 f_1^2}+\frac{P_2 f_1}{r_2^2 f_2^3}\right)\frac{1}{\left(1+\frac{P_1}{f_1}+\frac{P_2}{f_2}\right)^3} \tag{5.66}$$

$$\frac{d^2 f_2}{d\xi^2}=\frac{1}{r_2^4 f_2^3}-2\beta_1\left(\frac{P_1 f_2}{r_1^2 f_1^3}+\frac{P_2}{r_2^2 f_2^2}\right)\frac{1}{\left(1+\frac{P_1}{f_1}+\frac{P_2}{f_2}\right)^2}$$
$$-4\beta_2\left(\frac{P_1 f_2}{r_1^2 f_1^3}+\frac{P_2}{r_2^2 f_2^2}\right)\frac{1}{\left(1+\frac{P_1}{f_1}+\frac{P_2}{f_2}\right)^3} \tag{5.67}$$

The change in spatial width of both the solitons with propagation can be inferred from (5.66) and (5.67). For a stable soliton pair, it is necessary to look for non-diverging solutions. Hence, $\frac{d^2 f_1}{d\xi^2}=\frac{d^2 f_2}{d\xi^2}=0$ for points of equilibrium of (5.66) and (5.67) and so,

$$\frac{1}{r_1^4}-2\beta_1\left(\frac{P_1}{r_1^2}+\frac{P_2}{r_2^2}\right)\frac{1}{(1+P_1+P_2)^2}-4\beta_2\left(\frac{P_1}{r_1^2}+\frac{P_2}{r_2^2}\right)\frac{1}{(1+P_1+P_2)^3}=0 \tag{5.68}$$

$$\frac{1}{r_2^4}-2\beta_1\left(\frac{P_1}{r_1^2}+\frac{P_2}{r_2^2}\right)\frac{1}{(1+P_1+P_2)^2}-4\beta_2\left(\frac{P_1}{r_1^2}+\frac{P_2}{r_2^2}\right)\frac{1}{(1+P_1+P_2)^3}=0 \tag{5.69}$$

Since (5.68) and (5.69) are coupled equations, we shall look for common solutions by equating the LHS of both equations. This in turn implies that $r_1=r_2$ from which we can infer that each soliton in the coupled soliton pair has identical spatial width. An important observation here is that spatial width of each soliton in the coupled soliton pair should be same for stable propagation. Putting $r_1=r_2=r$ in (5.68) and (5.69), we obtain,

$$\frac{1}{r^2}=\frac{2\beta_1(P_1+P_2)}{(1+P_1+P_2)^2}+\frac{4\beta_2(P_1+P_2)}{(1+P_1+P_2)^3} \tag{5.70}$$

The Eq. (5.56) validates the existence of a bright incoherently coupled soliton pair in photorefractive materials. To keep r positive, LHS also has to be positive.

As an example, consider the PMN-0.33PT crystal. Since we consider a photorefractive crystal exhibiting both the linear and quadratic electro-optic effect simultaneously, $a=1$, $b=1$. The parameters used in the calculation are shown in Table 5.4.

The existence curves for the coupled soliton pairs have been plotted in Figs. 5.6 and 5.7 for both, low and high powers respectively. Stably propagating coupled soliton pairs with a given peak power and spatial width exist for peak power values of each soliton component lying on the existence curve for the respective r. From the existence curves, it can be inferred that the power of any one soliton component

Table 5.4 Parameters considered for PMN-0.33PT crystal (Reprinted by permission from Springer Nature: Springer Applied Physics B: Lasers and Optics, Incoherently coupled Gaussian soliton pairs in biased photorefractive crystal having both the linear and quadratic electro-optic effect, Aavishkar Katti, Copyright 2018) [32–35]

Parameter	Value	Parameter	Value
n_e	2.562	l (*crystal width*)	1 cm
x_0	20 μm	V (*bias emf*)	1000 V
λ_0	632.8 nm	β_1	15.4612
r_{eff}	182×10^{-12} m/V	β_2	1.1724
$g_{eff}\epsilon_0^2(\epsilon_r - 1)^2$	$1.38 \times 10^{-16} \text{m}^2/\text{V}^2$	ρ	0

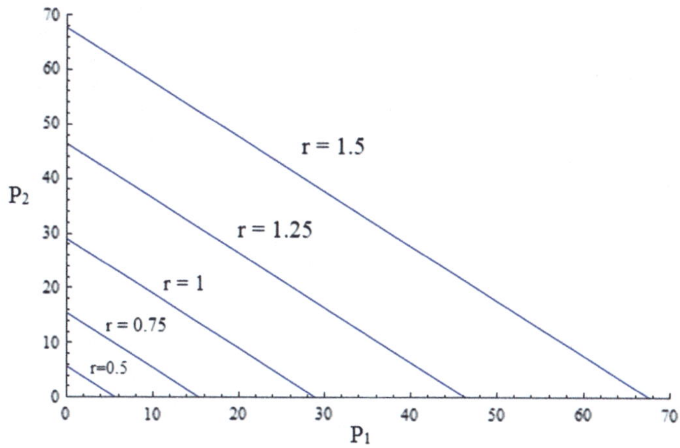

Fig. 5.7 The plot of peak power of one component versus the peak power of the other signifying the existence curve for the bright coupled soliton pairs for high powers. (Reprinted by permission from Springer Nature: Springer Applied Physics B: Lasers and Optics, Incoherently coupled Gaussian soliton pairs in biased photorefractive crystal having both the linear and quadratic electro-optic effect, Aavishkar Katti, Copyright 2018)

can be much lesser than that of the other soliton component. Hence, a light beam of a particular spatial width can be self trapped by making another strong optical beam to propagate collinearly with it (Fig. 5.8).

Figures 5.9 and 5.10 plot the spatial soliton width versus any one soliton component's peak power with the other component's peak power fixed for low and high powers respectively.

If both constituent components of the coupled soliton pair have equal power, i.e., $P_1 = P_2 = P$, we get, from (5.28),

$$\frac{1}{r^2} = \frac{4\beta_1 P}{(1+2P)^2} + \frac{8\beta_2 P}{(1+2P)^3} \tag{5.71}$$

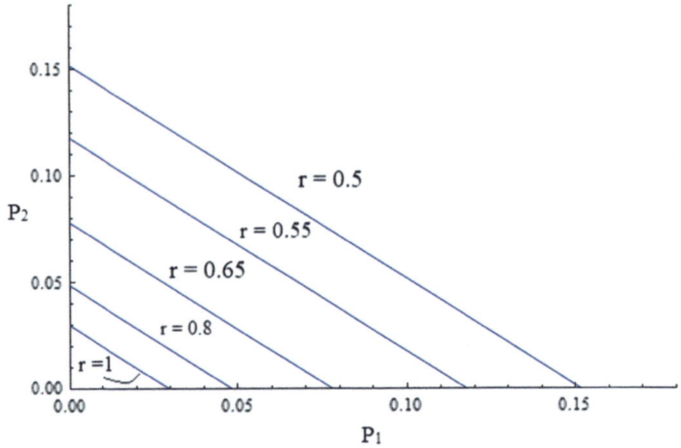

Fig. 5.8 The plot of peak power of one component versus the peak power of the other signifying the existence curve for the bright coupled soliton pairs for low powers. (Reprinted by permission from Springer Nature: Springer Applied Physics B: Lasers and Optics, Incoherently coupled Gaussian soliton pairs in biased photorefractive crystal having both the linear and quadratic electro-optic effect, Aavishkar Katti, Copyright 2018)

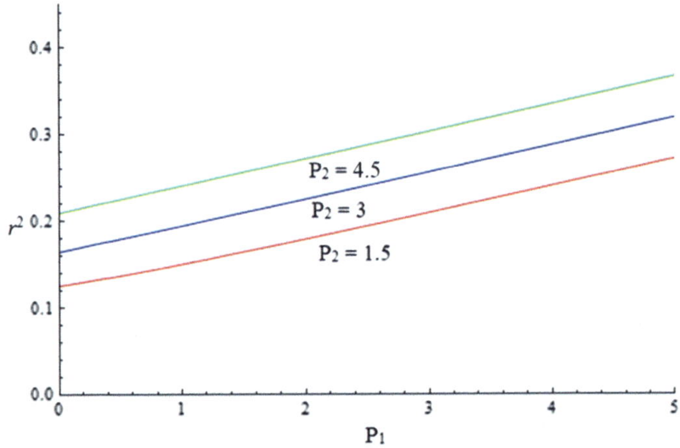

Fig. 5.9 Variation of the gaussian soliton pair width with the peak power of one component (P_1) when the peak power of the other (P_2) is constant (for high powers). (Reprinted by permission from Springer Nature: Springer Applied Physics B: Lasers and Optics, Incoherently coupled Gaussian soliton pairs in biased photorefractive crystal having both the linear and quadratic electro-optic effect, Aavishkar Katti, Copyright 2018)

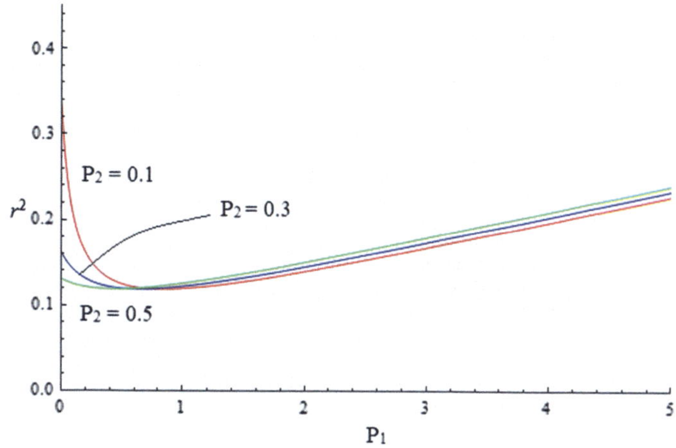

Fig. 5.10 Variation of the gaussian soliton pair width with the peak power of one component (P_1) when the peak power of the other (P_2) is constant(for low powers). (Reprinted by permission from Springer Nature: Springer Applied Physics B: Lasers and Optics, Incoherently coupled Gaussian soliton pairs in biased photorefractive crystal having both the linear and quadratic electro-optic effect, Aavishkar Katti, Copyright 2018)

The spatial width of the Gaussian soliton pair as a function of the peak power is shown Fig. 5.11. Bistable states can been observed very clearly, i.e., two distinct coupled soliton pairs can exist with dissimilar peak power but identical spatial width.

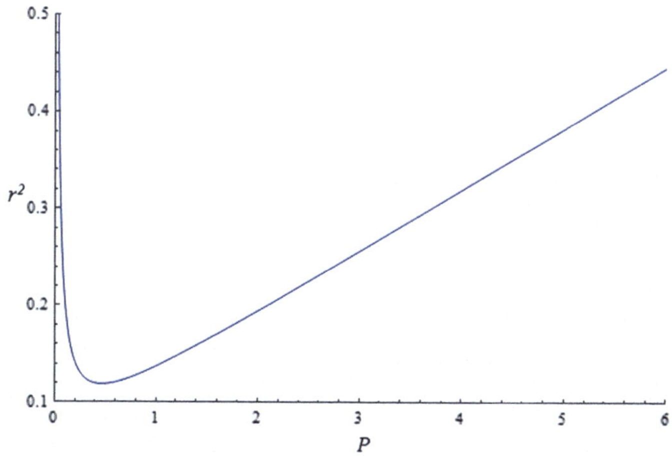

Fig. 5.11 Gaussian soliton pair's width as a function of peak power, under the condition $P_1 = P_2 = P$. (Reprinted by permission from Springer Nature: Springer Applied Physics B: Lasers and Optics, Incoherently coupled Gaussian soliton pairs in biased photorefractive crystal having both the linear and quadratic electro-optic effect, Aavishkar Katti, Copyright 2018)

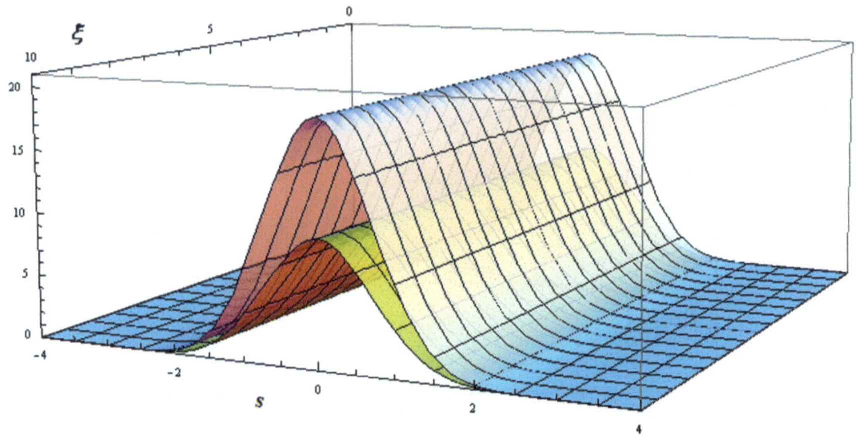

Fig. 5.12 Propagation of the coupled soliton pair with normalized distance ξ. $P_1 = 10.00$ and $P_2 = 19.04$ and $r = 1$. (Reprinted by permission from Springer Nature: Springer Applied Physics B: Lasers and Optics, Incoherently coupled Gaussian soliton pairs in biased photorefractive crystal having both the linear and quadratic electro-optic effect, Aavishkar Katti, Copyright 2018)

For investigating the propagation of these incoherently coupled quasi-solitons, an arbitrary value of r is chosen. The value of r corresponds to a unique existence curve for P_1 and P_2. If any point on the existence curve is selected, Eqs. (5.66) and (5.67) can be solved numerically with the particular P_1 and P_2 values for f_1 and f_2 with the initial conditions $f_1(0) = f_2(0) = 1$ and $\frac{df_1}{d\xi} = \frac{df_2}{d\xi} = 0$ at $\xi = 0$. Of course, this implies that f_1 and f_2 remain constant during propagation. The propagation of the bright incoherently coupled gaussian soliton pair with normalized distance is shown in Fig. 5.12.

5.6 Concluding Remarks and Further Reading

We have seen a comprehensive theoretical formulation for coupled solitons in different types of photorefractive media including photovoltaic and non-photovoltaic photorefractive media with linear or quadratic nonlinearites. Coupled soliton pairs can be either incoherently coupled or coherently coupled depending upon the mutual phase relationship between the two light beams and both of these are studied in detail. A unique type of coupled soliton pair, known as Gaussian or quasi soliton pair have also been illustrated which belong to new parameter space and are quite different in characteristics from the conventional coupled soliton pairs.

The reader is referred to Ref. [9] where an extension of the theory to incorporate coupled multi component solitons has been undertaken. The analysis in Sects. 5.2 or 5.3 carries forward exactly but the soliton components' ansatz are now taken as extended theta projections. References [4, 6, 7, 9, 11, 12, 14, 36–39] provide a thorough study of incoherently coupled spatial soliton pairs in diverse configurations of photorefractive crystals. Similarly, Refs. [40–43] study the coherent coupling of solitons in different configurations of photorefractive media.

In case of Gaussian soliton pairs, the fact that existence of solitons allows for an infinite choice of each soliton component's power is a very important distinguishing property. References [18, 44–46] deal with the theoretical formulation for these unique type of soliton pairs in diverse photorefractive systems.

References

1. R.Y. Chiao, E. Garmire, C.H. Townes, Self-trapping of optical beams. Phys. Rev. Lett. **13**(15), 479 (1964)
2. R. De La Fuente, A. Barthelemy, Spatial solitons pairing by cross phase modulation. Opt. Commun. **88**(4–6), 419–423 (1992)
3. M. Shalaby, A.C. Barthelemy, Observation of the self-guided propagation of a dark and bright spatial soliton pair in a focusing nonlinear medium. IEEE J. Quantum Electron. **28**(12), 2736–2741 (1992)
4. Z. Chen, M. Segev, T.H. Coskun, D.N. Christodoulides, Y.S. Kivshar, Coupled photorefractive spatial-soliton pairs. J. Opt. Soc. Am. B **14**(11), 3066 (1997). https://doi.org/10.1364/JOSAB.14.003066
5. Z. Chen, M. Segev, T.H. Coskun, D.N. Christodoulides, Observation of incoherently coupled photorefractive spatial soliton pairs. Opt. Lett. **21**(18), 1436–1438 (1996)
6. Z. Chen, M. Segev, T.H. Coskun, D.N. Christodoulides, Y.S. Kivshar, V.V. Afanasjev, Incoherently coupled dark-bright photorefractive solitons. Opt. Lett. **21**(22), 1821–1823 (1996). https://doi.org/10.1364/OL.21.001821
7. H. Chun-feng et al., Incoherently coupled screening-photovoltaic soliton families in biased photovoltaic photorefractive crystals. Chin. Phys. **10**(4), 310–313 (2001). https://doi.org/10.1088/1009-1963/10/4/309
8. D.N. Christodoulides, M.I. Carvalho, Bright, dark, and gray spatial soliton states in photorefractive media. J. Opt. Soc. Am. B **12**(9), 1628 (1995). https://doi.org/10.1364/JOSAB.12.001628
9. L. Keqing, Z. Yanpeng, T. Tiantong, L. Bo, Incoherently coupled steady-state soliton pairs in biased photorefractive-photovoltaic materials. Phys. Rev. E Stat., Nonlin. Soft Matter Phys. **64**(5) II, 056603/1–056603/9 (2001). https://doi.org/10.1103/PhysRevE.64.056603
10. P. Günter, J. Huignard, A. Glass, *Photorefractive Materials and Their Applications 1* (Springer Science+Business Media Inc., New York, 2006).
11. A. Katti, R.A. Yadav, D.P. Singh, Theoretical investigation of incoherently coupled solitons in centrosymmetric photorefractive crystals. Opt. Int. J. Light Electron Opt. **136**, 89–106 (2017). https://doi.org/10.1016/j.ijleo.2017.01.099
12. L. Hao, Q. Wang, C. Hou, Incoherently coupled spatial soliton pairs due to both the linear and quadratic electro-optic effects. J. Mod. Opt. **62**(3), 231–237 (2015). https://doi.org/10.1080/09500340.2014.968644
13. J.S. Liu, K.Q. Lu, Screening-photovoltaic spatial solitons in biased photovoltaic-photorefractive crystals and their self-deflection. J. Opt. Soc. Am. B-Opt. Phys. **16**(4), 550–555 (1999)
14. L. Hao, Q. Wang, C. Hou, Incoherently coupled spatial soliton pairs due to both the linear and quadratic electro-optic effects. J. Mod. Opt. **62**(3), 205–211 (2015). https://doi.org/10.1080/09500340.2014.968644
15. L. Hao, C. Hou, Q. Wang, H. Mu, Coherently coupled spatial soliton pairs in biased photorefractive crystals with both the linear and quadratic electro-optic effects. Optik **127**(10), 4339–4344 (2016)

16. S. Konar, S. Jana, S. Shwetanshumala, Incoherently coupled screening photovoltaic spatial solitons in biased photovoltaic photorefractive crystals. Opt. Commun. **273**(2), 324–333 (2007). https://doi.org/10.1016/J.OPTCOM.2007.01.051
17. S. Shwetanshumala, S. Konar, Bright optical spatial solitons in a photorefractive waveguide. Phys. Scripta **82**(4), 045404 (2010). https://doi.org/10.1088/0031-8949/82/04/045404
18. A. Katti, Incoherently coupled Gaussian soliton pairs in biased photorefractive crystal having both the linear and quadratic electro-optic effect. Appl. Phys. B **124**(9), 192 (2018)
19. A. Katti, Bright pyroelectric quasi-solitons in a photorefractive waveguide. Optik **156**, 433–438 (2018). https://doi.org/10.1016/j.ijleo.2017.10.105
20. S. Konar, N. Asif, External field-controlled photovoltaic symbiotic spatial soliton pairs in two-photon photorefractive crystals. Phys. Scripta **81**(1), 015401 (2009)
21. A. Katti, Bright screening solitons in a photorefractive waveguide. Opt. Quant Electron. **50**(6), 263 (2018). https://doi.org/10.1007/s11082-018-1524-y
22. G.C. Valley, M. Segev, B. Crosignani, A. Yariv, M.M. Fejer, M.C. Bashaw, Dark and bright photovoltaic spatial solitons. Phys. Rev.s A **50**(6), R4457 (1994). https://doi.org/10.1103/PhysRevA.50.R4457
23. S.A. Akhmanov, A.P. Sukhoruk, Self-focusing and diffraction of light beams in a nonlinear medium. Sov. Phys. Uspekhi-USSR **10**(5), 609–636 (1968)
24. D. Anderson, Variational approach to nonlinear pulse propagation in optical fibers. Phys. Rev. A **27**(6), 3135–3145 (1983). https://doi.org/10.1103/PhysRevA.27.3135
25. S.N. Vlasov, V.A. Petrishchev, V.I. Talanov, Averaged description of wave beams in linear and nonlinear media (the method of moments). Radiophys. Quantum Electron. **14**(9), 1062–1070 (1974). https://doi.org/10.1007/BF01029467
26. V. Skarka, V.I. Berezhiani, R. Miklaszewski, Spatiotemporal soliton propagation in saturating nonlinear optical media. Phys. Rev. E Stat. Phys. Plasmas Fluids Related Interdiscip. Top. **56**(1), 1080–1087 (1997). https://doi.org/10.1103/PhysRevE.56.1080
27. Z. Liu, W. Zang, J. Tian, W. Zhou, C. Zhang, G. Zhang, Analysis of Z-scan of thick media with high-order nonlinearity by variational approach. Opt. Commun. **219**(1–6), 411–419 (2003). https://doi.org/10.1016/S0030-4018(03)01298-7
28. Y. Huang, Q. Guo, J. Cao, Optical beams in lossy non-local Kerr media. Opt. Commun. **261**(1), 175–180 (2006). https://doi.org/10.1016/j.optcom.2005.12.003
29. P. Tchofo Dinda, A.B. Moubissi, K. Nakkeeran, A collective variable approach for dispersion-managed solitons. J. Phys. A: Math. Gener. **34**(10), L103–L110 (2001). https://doi.org/10.1088/0305-4470/34/10/104
30. J.H.B. Nijhof, W. Forysiak, N.J. Doran, The averaging method for finding exactly periodic. IEEE J. Sel. Top. Quantum Electron. **6**(2), 330–336 (2000)
31. J.N. Kutz, S.D. Koehler, L. Leng, K. Bergman, Analytic study of orthogonally polarized solitons interacting in highly birefringent optical fibers. J. Opt. Soc. Am. B-Opt. Phys. **14**(3), 636–642 (1997)
32. X. Wan, D. Wang, X. Zhao, H. Luo, H.C.-Chan, State, and Undefined, *Electro-optic Characterization of Tetragonal (1−x)Pb(Mg1/3Nb2/3)*O_3*–xPbTiO*$_3$*Single Crystals* (Elsevier, 2005)
33. R. Zhang, B. Jiang, W. Cao, Elastic, piezoelectric, and dielectric properties of multidomain 0.67Pb(Mg1/3Nb2/3)O_3–0.33PbTiO_3 single crystals. J. Appl. Phys. **90**(7), 3471–3475 (2001). https://doi.org/10.1063/1.1390494
34. Q. Jiang, Y. Su, H. Nie, Z. Ma, Y. Li, Separate spatial soliton pairs in a biased series photorefractive crystal circuit with both the linear and quadratic electro-optic effects. J. Mod. Opt. **64**(6), 609–615 (2016)
35. L. Hao, Q. Wang, C. Hou, Spatial solitons in biased photorefractive materials with both the linear and quadratic electro-optic effects. J. Mod. Opt. **61**(15), 1236–1245 (2014). https://doi.org/10.1080/09500340.2014.928379
36. D.N. Christodoulides, S.R. Singh, M.I. Carvalho, M. Segev, Incoherently coupled soliton pairs in biased photorefractive crystals. Appl. Phys. Lett. **68**(13), 1763–1765 (1996). https://doi.org/10.1063/1.116659

37. C. Hou, Z.X. Zhou, X.D. Sun, B.H. Yuan, Incoherently coupled grey-grey screening-photovoltaic soliton pairs in biased photovoltaic-photorefractive crystals. Optik **112**(1), 17–20 (2001). https://doi.org/10.1078/0030-4026-00003
38. B. Kumari, A. Katti, P.A. Alvi, Incoherently coupled grey solitons in photorefractive multiple quantum well planar waveguides. AIP Conf. Proc. **2100**(1), 020135 (2019). https://doi.org/10.1063/1.5098689
39. Y. Su, Q. Jiang, X. Ji, J. Wang, Incoherently coupled spatial soliton families in photorefractive polymers. Opt. Lasers Eng. **49**, 526–529 (2011). https://doi.org/10.1016/j.optlaseng.2011.01.001
40. L. Hao, C. Hou, X. Wang, Q. Wang, H. Mu, Coherently coupled bright-bright screening soliton pairs in biased centrosymmetric photorefractive crystals. Opt.-Int. J. **127**(15), 5928–5934 (2016)
41. A. Katti, Coherently coupled solitons in photorefractive media due to pyroelectric effect. J. Nonlinear Opt. Phys. Mater. **26**(04), 1750044 (2017). https://doi.org/10.1142/S0218863517500448
42. A. Katti, *Coherently Coupled Pyroelectric Photovoltaic Solitons in Photovoltaic Photorefractive Media*," no. September, 2016, [Online]. Available: https://www.researchgate.net/profile/Aavishkar_Katti2/publication/308294281_COHERENTLY_COUPLED_PYROELECTRIC_PHOTOVOLTAIC_SOLITONS_IN_PHOTOREFRACTIVE_MEDIA/links/57dfc1f108ae1dcfea8578d9.pdf
43. B. Kumari, A. Katti, P.A. Alvi, Coupling of optical spatial solitons in photorefractive multiple quantum well planar waveguide. Optik **183**, 1048–1060 (2019). https://doi.org/10.1016/j.ijleo.2019.02.106
44. A. Katti, Gaussian soliton pairs in an unbiased photorefractive crystal due to the pyroelectric effect. Eur. Phys. J. Plus **134**(12), 621 (2019)
45. K. Zhan, C. Hou, H. Tian, Y. Zhang, Incoherently coupled Gaussian soliton pairs in biased centrosymmetric photorefractive crystals. Opt. Laser Technol. **42**(7), 1176–1179 (2010). https://doi.org/10.1016/j.optlastec.2010.03.006
46. B.P. Akhouri, P.K. Gupta, Waveguiding effect on optical spatial solitons in centrosymmetric photorefractive materials. J. Opt. **46**(3), 281–286 (2017). https://doi.org/10.1007/s12596-016-0372-z

Chapter 6
Photorefractive Crystal Circuits

6.1 Introduction

We have seen, in the previous chapter about coupling of optical spatial solitons in photorefractive crystals. Various types of coupled soliton pairs and coupled soliton families can be observed in photorefractive crystals, viz. bright, dark, bright-dark, grey and even grey-bright and grey-dark. The mechanism of such coupling of soliton is abundantly clear as being due to a formation of an effective index waveguide due to both components of a soliton pair which then guides the coupled soliton pair [1, 2].

In contrast to this, a phenomenon known as separate coupling is quite interesting and has been studied quite extensively recently [3–7]. A circuit is considered which consists of two photorefractive crystals connected in series. The circuit may be biased or unbiased. A light beam is made to propagate along each crystal and the self trapping of the light beams in each crystal is studied. The soliton in each crystal induces a current, and this light induced current can flow into the other crystal and hence the solitons in both crystals can affect each other. The spatial profiles of the solitons, their dynamical evolution and stabilities are affected by each other [5]. Normally, non photovoltaic photorefractive crystal circuits have to be biased with an external voltage while photovoltaic crystal circuits need not be biased as photovoltaic crystals themselves act as a current source.

In this chapter, we shall first formulate a general theory for finding out the numerical solutions proving the existence of such separate coupled solitons in different types of configurations of photorefractive crystals. The approach to deriving the coupled space charge fields in both the crystals of the photorefractive crystal circuit will be discussed. The theoretical formulation for different types of separately coupled solitons will be elucidated. Taking a specific case of a centrosymmetric photorefractive crystal circuit, the coupling effect of input intensity and the crystal temperature on the solitons in both crystals of the circuit will be studied. The reader shall be given a brief idea and referred to suitable literature for further reading about the effect on dynamical evolutions and stabilities of the separately coupled solitons.

A. Katti and R. A. Yadav, *Optical Spatial Solitons in Photorefractive Materials*, Progress in Optical Science and Photonics 14, https://doi.org/10.1007/978-981-16-2550-3_6

6.2 Theoretical Foundations

To study the separate spatial solitons in a biased photorefractive crystal circuit, we know that there exists a self trapped soliton beam in each crystal. For different orientations of the external bias field with the crystal optical axis, we can find the corresponding circumstances to form bright or dark solitons in the individual crystals and hence, the bright, dark or grey or bright-dark separately coupled soliton pairs. We illustrate this in Fig. 6.1. For example, if we consider conventional photorefractive crystals with the linear electro-optic effect, then in Fig. 6.1, we can clearly see that both the crystals can support dark solitons if c-axis is parallel to the x-axis and $\hat{c}$-axis is anti-parallel to the $\hat{x}$ axis. Both crystals can support bright solitons if the c-axis is anti-parallel to the x-axis and $\hat{c}$-axis is parallel to the $\hat{x}$ axis. And finally, one crystal supports a bright soliton while the other supports a dark soliton if the c-axis is parallel to the x-axis and $\hat{c}$-axis is parallel to the $\hat{x}$ axis. Similar considerations can be applied to photorefractive crystals having the quadratic electro-optic effect or exhibiting both the linear and quadratic electro-optic effect simultaneously. We shall be discussing these eventually. For now, we shall assume that the two conventional photorefractive crystals have a soliton beam propagating in each. The intensities of the two beams are I and $\hat{I}$. The propagation direction is labeled to be the z and $\hat{z}$ axes and we shall assume diffraction of the two individual beams to be along the x and $\hat{x}$ directions. The two photorefractive crystals will be designated by P and $\hat{P}$. In continuation, $\hat{.}$ denotes the parameters in the crystal $\hat{P}$. The c-axis of the photorefractive crystals P and $\hat{P}$ are kept either parallel or anti parallel to the x axis and $\hat{x}$ axis. The direction of c-axis depends on which type of soliton is supported by the crystal. For instance, considering a photorefractive crystal with the linear electro-optic effect, the c-axis parallel to the

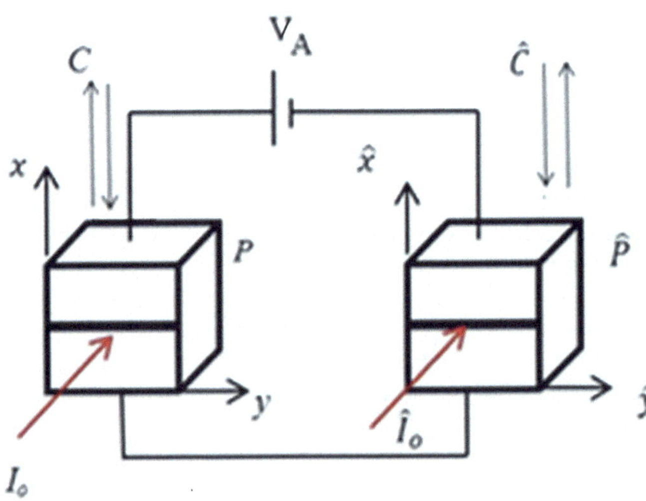

Fig. 6.1 Circuit for realization of separately coupled solitons in biased photorefractive crystal series circuits

x-axis and antiparallel to the electric field direction supports the formation of a dark or grey soliton while a c-axis antiparallel to the x-axis and parallel to the electric field supports a bright soliton. In contrast, for a centrosymmetric photorefractive crystal, the parallel or antiparallel nature of the c-axis with the x-axis and electric field does not matter since we now have a quadratic electro-optic effect and the change in refractive index depends upon the square of the space charge field. Similar considerations work for photorefractive crystals having both quadratic and linear electro-optic effect simultaneously and photovoltaic photorefractive crystals and will be discussed eventually. The electric field envelopes in the respective crystals are expressed as, $E_1 = \hat{x}\varphi(x,z)\exp(ikz)$, $E_2 = \widehat{\hat{x}}\varphi(\hat{x},\hat{z})\exp(ik\hat{z})$. The dynamical evolution equation for the soliton in the crystal P is,

$$\left(i\frac{\partial}{\partial z} + \frac{1}{2k}\frac{\partial^2}{\partial x^2} + \frac{k}{n_e}\Delta n\right)\varphi(x,z) = 0 \tag{6.1}$$

$$\Delta n_e = -a\frac{n_e^3 r_{33} E_{sc}}{2} - b\frac{n_e^3 g_{eff}\epsilon_0^2(\epsilon_r - 1)^2 E_{sc}^2}{2} \tag{6.2}$$

a and b are constants which can be either individually zero or one or both depending upon whether the photorefractive crystal has the linear electro-optic effect, quadratic electro-optic effect or exhibits both electro-optic effects simultaneously. r_{33} is the linear electro-optic coefficient and g_{eff} is the quadratic electro-optic coefficient. Other symbols have their usual meaning. The space charge field E_{sc} can be derived using the charge transport equations exactly following the approach discussed in Chap. 1,

$$\gamma_R n N_D^+ = s_i(I + I_b + I_d)(N_D - N_D^+) \tag{6.3}$$

$$\frac{\partial E_{sc}}{\partial x} = \frac{e}{\varepsilon_0\varepsilon}(N_D^+ - N_A - n) \tag{6.4}$$

$$J = e\mu n E_{sc} + k_B T\mu\frac{\partial n}{\partial x} \tag{6.5}$$

$$\frac{\partial J}{\partial x} = 0 \tag{6.6}$$

where I_b is the background intensity and other symbols have their usual meaning as defined before. In a photorefractive crystal, we work under the approximation $N_D^+ \approx N_A$ and hence, from (6.3),

$$n = \frac{s_i(N_D - N_A)}{\gamma_R N_A}(I + I_b + I_d) \tag{6.7}$$

The electron density in regions of uniform illumination $x \to \pm\infty$ is, from (6.7),

$$n_\infty = \frac{s_i(N_D - N_A)}{\gamma_R N_A}(I_\infty + I_b + I_d) \tag{6.8}$$

From (6.6), we get to know that the current density is constant, and hence, $J_\infty(x \to \pm\infty, z) = J$.

So from (6.5),

$$J_\infty = e\mu n_\infty E_0 \tag{6.9}$$

where, as usual E_0 is the electric field in regions of uniform illumination $E_0 = E_{sc}(x \to \pm\infty)$.

Now, we can derive these same relations for the crystal $\hat{P}$,

$$\hat{n}_\infty = \frac{\hat{s}_i\left(\hat{N}_D - \hat{N}_A\right)}{\hat{\gamma}_R \hat{N}_A}\left(\hat{I}_\infty + \hat{I}_b + \hat{I}_d\right) \tag{6.10}$$

$$\hat{J}_\infty = \hat{e}\hat{\mu}\hat{n}_\infty \hat{E}_0 \tag{6.11}$$

$V, \hat{V}$ are the voltages measured between the electrodes of the crystals $P, \hat{P}$ respectively having width $W, \hat{W}$. If $S, \hat{S}$ denote the surface area of the electrodes at crystal $P, \hat{P}$ respectively, then $SJ = \hat{S}\hat{J}$ since the crystals are connected in series. The soliton beam's spatial extent is assumed to be much smaller than the x-width of the crystals and hence we have, $E_{sc}(x \to \infty) = E_o = V/W$ and $\hat{E}_{sc}(\hat{x} \to \infty) = \hat{E}_0 = \hat{V}/\hat{W}$. Here again, since the two crystals are connected in series, $V + \hat{V} = V_A$ where V_A the source voltage. From these aforesaid conditions and using (6.10) and (6.11), we can deduce the expressions for $\hat{E}_0$ and E_0,

$$E_0 = gE_A \tag{6.12}$$

$$\hat{E}_0 = \hat{g}\hat{E}_A \tag{6.13}$$

with,

$$g = \frac{\hat{\delta}\left(\hat{I}_\infty + \hat{I}_b + \hat{I}_d\right)}{\delta(I_\infty + I_b + I_d) + \hat{\delta}\left(\hat{I}_\infty + \hat{I}_b + \hat{I}_d\right)} \tag{6.14}$$

$$\hat{g} = \frac{\delta(I_\infty + I_b + I_d)}{\delta(I_\infty + I_b + I_d) + \hat{\delta}\left(\hat{I}_\infty + \hat{I}_b + \hat{I}_d\right)} \tag{6.15}$$

where, $\delta = S\mu s_i(N_D - N_A)/(\gamma_R N_A W)$, $E_A = V_A/W$, $I_\infty = I(x \to \infty)$, $\hat{\delta} = \hat{S}\hat{\mu}\hat{s}_i\left(\hat{N}_D - \hat{N}_A\right)/\left(\hat{\gamma}_R \hat{N}_A \hat{W}\right)$, $\hat{E}_A = V_A/\hat{W}$, $\hat{I}_\infty = \hat{I}(x \to \infty)$, g and $\hat{g}$ are known as coupling coefficients. It is interesting to note that $\hat{g} + g = 1$ indicating a perfect coupling of the space charge field of the two crystals. As $J_\infty(x \to \infty, z) = J$,

$$E_{sc} = E_0 \frac{I_\infty + I_b + I_d}{I + I_b + I_d} = g E_A \frac{I_\infty + I_b + I_d}{I + I_b + I_d} \tag{6.16}$$

$$\hat{E}_{sc} = \hat{E}_0 \frac{\hat{I}_\infty + \hat{I}_b + \hat{I}_d}{\hat{I} + \hat{I}_b + \hat{I}_d} = \hat{g} \hat{E}_A \frac{\hat{I}_\infty + \hat{I}_b + \hat{I}_d}{\hat{I} + \hat{I}_b + \hat{I}_d} \tag{6.17}$$

We can see clearly from Eqs. (6.16) and (6.17) that the space charge fields are coupled to each other. Putting (6.16) into (6.2) and (6.1), we can obtain the evolution equation for the soliton in the crystal P,

$$i\frac{\partial U}{\partial \xi} + \frac{1}{2}\frac{\partial^2 U}{\partial s^2} - \beta_1(1+\rho)\frac{1}{1+|U|^2}U - \beta_2(1+\rho)^2\frac{1}{(1+|U|^2)^2}U = 0 \tag{6.18}$$

Similarly, using (6.17), we can derive an evolution equation for the soliton in crystal $\hat{P}$,

$$i\frac{\partial \hat{U}}{\partial \xi} + \frac{1}{2}\frac{\partial^2 \hat{U}}{\partial \hat{s}^2} - \hat{\beta}_1(1+\hat{\rho})\frac{1}{1+|\hat{U}|^2}\hat{U} - \hat{\beta}_2(1+\hat{\rho})^2\frac{1}{(1+|\hat{U}|^2)^2}\hat{U} = 0 \tag{6.19}$$

where,

$U = [2\eta_0(I_b + I_d)/n_e]^{-1/2}\varphi$, $\rho = I_\infty/(I_d + I_b)$, $\beta_1 = a(k_0x_0)^2 n_e^4 r_{33} g E_A/2$, $\beta_2 = b(k_0x_0)^2 n_e^4 g_{eff}\epsilon_0^2(\epsilon_r - 1)^2(gE_A)^2/2$, $s = x/x_0$, $\xi = z/(kx_0)^2$.

and,

$\hat{U} = [2\eta_0\left(\hat{I}_b + \hat{I}_d\right)/n_e]^{-1/2}\varphi$, $\hat{\rho} = \hat{I}_\infty/\left(\hat{I}_d + \hat{I}_b\right)$, $\hat{\beta}_1 = \hat{a}(\hat{k}_0\hat{x}_0)^2\hat{n}_e^4\hat{r}_{33}\hat{g}E_A/2$, $\hat{\beta}_2 = \hat{b}(\hat{k}_0\hat{x}_0)^2\hat{n}_e^4\hat{g}_{eff}\epsilon_0^2(\epsilon_r - 1)^2(\hat{g}E_A)^2/2$, $\hat{s} = \hat{x}/\hat{x}_0$, $\hat{\xi} = \hat{z}/(k\hat{x}_0)^2$.

6.3 Biased Series Centrosymmetric Photorefractive Crystal Circuit

As an illustration of theoretical model as described in Sect. 6.2, consider a case of separate coupling of grey spatial solitons in a biased centrosymmetric photorefractive crystal circuit [3]. If we consider the Fig. 6.1, we can have any orientation of the c-axis for both crystals since the refractive index change is now dependent upon the square of the space charge field. Since we consider centrosymmetric photorefractive crystals, i.e., having only the quadratic electro-optic effect, so we take $a = 0$ and $b = 1$. For such a photorefractive crystal circuit, we have the following dynamical evolution equations in crystal P and $\hat{P}$ from (6.18) and (6.19),

$$i\frac{\partial U}{\partial \xi} + \frac{1}{2}\frac{\partial^2 U}{\partial s^2} - \beta_2(1+\rho)^2\frac{U}{(1+|U|^2)^2} = 0 \tag{6.20}$$

$$i\frac{\partial \hat{U}}{\partial \hat{\xi}} + \frac{1}{2}\frac{\partial^2 \hat{U}}{\partial \hat{s}^2} - \hat{\beta}_2(1+\hat{\rho})^2\frac{\hat{U}}{(1+|\hat{U}|^2)^2} = 0 \tag{6.21}$$

Here, the intensity scales as $I = (I_b + I_d)|U|^2$. The total intensity $\hat{I} = \left(\hat{I}_b + \hat{I}_d\right)|\hat{U}|^2$. I_b denotes the background intensity and I_d denotes the dark irradiance. The expressions for $\hat{E}_0$ and E_0 are as given in (6.12) and (6.13) and the space charge field is as given in (6.16) and (6.17) along with g and $\hat{g}$ are the coupling coefficients in (6.14) and (6.15) satisfying $\hat{g} + g = 1$.

6.3.1 Grey-Grey Separate Soliton Pair

Expressing the grey soliton solution in crystal P as,

$$U = \rho^{1/2} y(s)\exp\left[i\left(c\xi + \int_0^s \frac{Qd\tilde{s}}{y^2(\tilde{s})}\right)\right] \tag{6.22}$$

In (6.22), Q is a constant while $y(s)$ is a real and bounded function such that it lies between 0 and 1. The boundary conditions for $y(s)$ would those applicable to grey solitons,

$$y(s \to \pm\infty) = 1,\ \dot{y}(0) = 0,\ y^2(0) = m(0 < m < 1),\ \ddot{y}(\infty) = 0,\ \dot{y}(\infty) = 0$$

Putting (6.22) in (6.20), we find the differential equation satisfied by $y(s)$,

$$\ddot{y} - 2cy - \frac{Q^2}{y^3} - 2\beta_2(1+\rho)^2\frac{y}{(1+\rho y^2)^2} = 0 \tag{6.23}$$

Using the boundary conditions at infinity, in (6.23), we can obtain the value of Q,

$$Q^2 = -2c - 2\beta_2 \tag{6.24}$$

Now, integrating (6.23) once, we get

$$\dot{y}^2 = 2c\left(y^2\right) - Q^2\left(\frac{1}{y^2}\right) - \frac{2\beta_2(1+\rho)^2}{\rho}\left(\frac{1}{1+\rho y^2}\right) + d \tag{6.25}$$

where d is a constant of integration. We have two constants we need to find, c and d. These can be found out by using the boundary conditions of $y(s)$ at $s = 0$ and $s \to \infty$ in (6.23) and (6.25).

Taking $y = y(0)$ in (6.25), and remembering the fact that $y(0)^2 = m$ and $\dot{y}(0) = 0$ gives,

$$2cm - \frac{Q^2}{m} - 2\beta_2 \frac{(1+\rho)^2}{\rho}\left(\frac{1}{1+\rho m}\right) + d = 0 \tag{6.26}$$

Again, considering the boundary conditions at infinity in (6.25),

$$2c - Q^2 - 2\beta_2 \frac{(1+\rho)}{\rho} + d = 0 \tag{6.27}$$

Solving (6.26) and (6.27),

$$c = \frac{\frac{-2\beta_2}{m} + 2\beta_2 + \frac{2\beta_2(1+\rho)^2}{\rho(1+\rho m}) - 2\beta_2\frac{1+\rho}{\rho}}{2(m-1) + \frac{2}{m} - 2} \tag{6.28}$$

$$d = \frac{\frac{Q^2}{m} - Q^2 m - \frac{2\beta_2(1+\rho)m}{\rho} + \frac{2\beta_2(1+\rho)^2}{\rho(1+\rho m)}}{1-m} \tag{6.29}$$

Finally, the soliton envelope can be obtained by integrating (6.25) once again, and using (6.28) and (6.29),

$$s = \pm \int_{y}^{\sqrt{m}} \left(2c\left(\tilde{y}^2\right) - Q^2\left(\frac{1}{\tilde{y}^2}\right) - \frac{2\beta_2(1+\rho)^2}{\rho}\left(\frac{1}{1+\rho\tilde{y}^2}\right) + d\right)^{-1/2} d\tilde{y} \tag{6.30}$$

We can easily infer from (6.25) that $\beta_2 < 0$ must be true for the realization of grey solitons. This is similar to the case of dark solitons.

We can perform an exactly similar derivation for the crystal $\hat{P}$ where we take the grey soliton solution as,

$$\hat{U} = \hat{\rho}^{1/2}\hat{y}(s)\exp\left[i\left(\hat{c}\xi + \int_0^s \frac{\hat{Q}d\tilde{\hat{s}}}{y^2\left(\tilde{\hat{s}}\right)}\right)\right] \tag{6.31}$$

and we get the soliton envelope as,

$$\hat{s} = \pm \int_{\hat{y}}^{\sqrt{\hat{m}}} \left(2\hat{c}\left(\tilde{\hat{y}}^2\right) - \hat{Q}^2\left(\frac{1}{\tilde{\hat{y}}^2}\right) - \frac{2\hat{\beta}_2(1+\hat{\rho})^2}{\hat{\rho}}\left(\frac{1}{1+\hat{\rho}\tilde{\hat{y}}^2}\right) + \hat{d}\right)^{-1/2} d\tilde{\hat{y}} \tag{6.32}$$

with,

$$\hat{c} = \frac{\frac{-2\hat{\beta}_2}{\hat{m}} + 2\hat{\beta}_2 + \frac{2\hat{\beta}_2(1+\hat{\rho})^2}{\hat{\rho}(1+\hat{\rho}\hat{m}}) - 2\hat{\beta}_2\frac{1+\hat{\rho}}{\hat{\rho}}}{2(\hat{m}-1) + \frac{2}{\hat{m}} - 2} \tag{6.33}$$

$$\hat{d} = \frac{\frac{\hat{Q}^2}{\hat{m}} - \hat{Q}^2\hat{m} - \frac{2\hat{\beta}_2(1+\hat{\rho})\hat{m}}{\hat{\rho}} + \frac{2\hat{\beta}_2(1+\hat{\rho})^2}{\hat{\rho}(1+\hat{\rho}\hat{m})}}{1-\hat{m}} \tag{6.34}$$

As previously found for the crystal P, $\hat{\beta}_2 < 0$ must be true for the existence of grey solitons in crystal $\hat{P}$. Here, it is apt to note that β_2 and $\hat{\beta}_2$ are a function of the coupling factors g and $\hat{g}$ and hence, the grey soliton in each crystal in the series centrosymmetric photorefractive crystal circuit is coupled to the other.

6.3.2 Bright-Grey Separate Soliton Pair

In this case, let us consider that a bright soliton propagates in the crystal P while a grey soliton propagates in the crystal $\hat{P}$. In crystal P, the beam envelope solution for the bright soliton can be expressed as,

$$U = r^{1/2} y(s) \exp(i\mu\xi) \tag{6.35}$$

where, $r = I(0)/(I_b + I_d)$ and $y(s)$ is a bounded function which satisfies $0 \le y(s) \le 1$ and the respective boundary conditions for bright solitons $\dot{y}(0) = 0$, $y(\pm\infty) = 0$, $\dot{y}(\pm\infty) = 0$, $y(0) = 1$, $\ddot{y}(\pm\infty) = 0$, while μ is propagation constant's nonlinear shift.

Substituting (6.35) in (6.20), we get,

$$\ddot{y} = 2\mu y + \frac{2\beta_2}{(1+ry^2)^2} y \tag{6.36}$$

Integrating (6.36) once, we get,

$$\dot{y}^2 = 2\mu y^2 - \frac{2\beta_2}{r}\left(\frac{1}{1+ry^2}\right) + c \tag{6.37}$$

where c is a constant of integration.

Evaluating the function $y(s)$ at $s \to \infty$ by using the boundary conditions $\dot{y}(\pm\infty) = 0$, $y(\pm\infty) = 0$ in (6.37), we get,

$$c = \frac{2\beta_2}{r} \tag{6.38}$$

Now, substituting $y = y(s = 0)$ and using the boundary conditions $y(0) = 1, \dot{y}(0) = 0$ in (6.37), we get,

$$\mu = -\frac{\beta_2}{1+r} \tag{6.39}$$

Putting the value of μ and c from (6.38) and (6.39) in (6.37), we get,

$$\dot{y}^2 = \frac{-2\beta_2(y^2)}{1+r} - \frac{2\beta_2}{r}\left(\frac{1}{1+ry^2}\right) + \frac{2\beta_2}{r} \tag{6.40}$$

We can get the soliton profile by integrating (6.40) once again,

$$\left(\frac{2\beta_2}{r}\right)^{1/2} s = \pm \int_{y}^{1} (1 - \frac{r\tilde{y}^2}{1+r} - \frac{1}{1+r\tilde{y}^2})^{-1/2} d\tilde{y} \tag{6.41}$$

We also infer from (6.37) that $\beta_2 > 0$ for existence of bright solitons.

Turning our attention towards the crystal $\hat{P}$, we need to assume a grey soliton ansatz,

$$\hat{U} = \hat{\rho}^{1/2}\hat{y}(s)\exp\left[i\left(\hat{c}\hat{\xi} + \int_{0}^{s} \frac{\hat{Q}d\tilde{\hat{s}}}{y^2\left(\tilde{\hat{s}}\right)}\right)\right] \tag{6.42}$$

The procedure for solving to obtain grey soliton solutions has been detailed in Sect. 6.3.1 where we have solved for a grey grey separately coupled soliton pair. Proceeding similarly by substituting (6.42) in (6.20) and solving, we get the soliton envelope,

$$\hat{s} = \pm \int_{\hat{y}}^{\sqrt{\hat{m}}} \left(2\hat{c}\left(\tilde{\hat{y}}^2\right) - \hat{Q}^2\left(\frac{1}{\tilde{\hat{y}}^2}\right) - \frac{2\hat{\beta}_2(1+\hat{\rho})^2}{\hat{\rho}}\left(\frac{1}{1+\hat{\rho}\tilde{\hat{y}}^2}\right) + \hat{d}\right)^{-1/2} d\tilde{\hat{y}} \tag{6.43}$$

with,

$$\hat{c} = \frac{\frac{-2\hat{\beta}_2}{\hat{m}} + 2\hat{\beta}_2 + \frac{2\hat{\beta}_2(1+\hat{\rho})^2}{\hat{\rho}(1+\hat{\rho}\hat{m})}) - 2\hat{\beta}_2\frac{1+\hat{\rho}}{\hat{\rho}}}{2(\hat{m}-1) + \frac{2}{\hat{m}} - 2} \tag{6.44}$$

$$\hat{d} = \frac{\frac{\hat{Q}^2}{\hat{m}} - \hat{Q}^2\hat{m} - \frac{2\hat{\beta}_2(1+\hat{\rho})\hat{m}}{\hat{\rho}} + \frac{2\hat{\beta}_2(1+\hat{\rho})^2}{\hat{\rho}(1+\hat{\rho}\hat{m})}}{1-\hat{m}} \tag{6.45}$$

in crystal $\hat{P}$, $\hat{\beta}_2 < 0$ for existence of grey solitons.

6.3.3 Grey-Dark Separate Soliton Pair

Consider a dark soliton propagating in crystal P. The dark soliton solution is expressed as,

$$U = \rho^{1/2} y(s) \exp(i\mu\xi) \tag{6.46}$$

where μ is a nonlinear shift of the propagation constant, $\rho = I_\infty/(I_b + I_d)$, and $y(s)$ is a bounded function such that $0 \le y(s) \le 1$. The boundary conditions satisfied by $y(s)$ will be $y(\pm\infty) = \pm 1$, $\dot{y}(\pm\infty) = 0$, $y(0) = 0$, $\ddot{y}(\pm\infty) = 0$.

Putting (6.46) in (6.20), we obtain,

$$\frac{1}{2}\ddot{y} - \mu y - \frac{\beta_2(1+\rho)^2}{(1+\rho y^2)^2} y = 0 \tag{6.47}$$

Integrating once, we get,

$$\frac{1}{2}\dot{y}^2 = \mu y^2 - \frac{\beta_2(1+\rho)^2}{\rho(1+\rho y^2)} + c \tag{6.48}$$

c is an integration constant. Using the boundary conditions at infinity in (6.47), we obtain,

$$\mu = -\beta_2 \tag{6.49}$$

Obtain the constant c by using the boundary conditions at infinity from (6.48) as,

$$c = \beta_2\left(\frac{2\rho+1}{\rho}\right) \tag{6.50}$$

Substituting c in (6.48), we have,

$$\dot{y}^2 = (-2\beta_2)\left(y^2 + \frac{(1+\rho)^2}{\rho(1+\rho y^2)} - \frac{2\rho+1}{\rho}\right) \tag{6.51}$$

from which we integrate once more to get,

$$(-2\beta_2)^{1/2} s = \int_y^0 \left[\left(\tilde{y}^2 + \frac{(1+\rho)^2}{\rho(1+\rho\tilde{y}^2)} - \frac{2\rho+1}{\rho}\right)\right]^{-1/2} d\tilde{y} \tag{6.52}$$

(6.52) gives us the spatial profile of the dark soliton. From (6.51), it can be inferred clearly that $\beta < 0$ must be satisfied for a dark soliton to form in P.

Now, considering a grey soliton propagating in crystal $\hat{P}$, the normalized field profile is,

$$\hat{U} = \hat{\rho}^{1/2}\hat{y}(s)\exp\left[i\left(\hat{c}\hat{\xi} + \int_{0}^{s} \frac{\hat{Q}d\tilde{\hat{s}}}{y^2\left(\tilde{\hat{s}}\right)}\right)\right] \tag{6.53}$$

The procedure for solving to obtain grey soliton solutions has been detailed in Sect. 6.3.1 where we have solved for a grey grey separately coupled soliton pair. Proceeding similarly by substituting (6.53) in (6.20) and solving, we get the soliton envelope,

$$\hat{s} = \pm \int_{\hat{y}}^{\sqrt{\hat{m}}} \left(2\hat{c}\left(\tilde{\hat{y}}^2\right) - \hat{Q}^2\left(\frac{1}{\tilde{\hat{y}}^2}\right) - \frac{2\hat{\beta}_2(1+\hat{\rho})^2}{\hat{\rho}}\left(\frac{1}{1+\hat{\rho}\tilde{\hat{y}}^2}\right) + \hat{d}\right)^{-1/2} d\tilde{\hat{y}} \tag{6.54}$$

with,

$$\hat{c} = \frac{\frac{-2\hat{\beta}_2}{\hat{m}} + 2\hat{\beta}_2 + \frac{2\hat{\beta}_2(1+\hat{\rho})^2}{\hat{\rho}(1+\hat{\rho}\hat{m}}) - 2\hat{\beta}_2\frac{1+\hat{\rho}}{\hat{\rho}}}{2\left(\hat{m}-1\right) + \frac{2}{\hat{m}} - 2} \tag{6.55}$$

$$\hat{d} = \frac{\frac{\hat{Q}^2}{\hat{m}} - \hat{Q}^2\hat{m} - \frac{2\hat{\beta}_2\left(1+\hat{\rho}\right)\hat{m}}{\hat{\rho}} + \frac{2\hat{\beta}_2(1+\hat{\rho})^2}{\hat{\rho}\left(1+\hat{\rho}\hat{m}\right)}}{1-\hat{m}} \tag{6.56}$$

It can be seen that $\hat{\beta}_2 < 0$ must be satisfied for a grey soliton to form in $\hat{P}$.

6.3.4 Intensity Effects on Separate Coupling

For illustration of the coupling effects, we assume the crystals to be both KLTN. There has to be one exception to consider here because of the condition $\beta_2 < 0$ for existence of grey and dark solitons. Hence, the value of $g_{\text{eff}} < 0$ will be taken for crystals supporting grey and dark solitons keeping all other parameters same.

For a grey-grey separate coupled soliton pair, we will consider the following parameters [3], $W = \hat{W} = 1cm.$, $g_{eff} = \hat{g}_{eff} = -0.12\,m^4/C^2$, $\delta = \hat{\delta}$, $I_d = \hat{I}_d$. Other parameters are taken, $\lambda_0 = \hat{\lambda}_0 = 0.5\,\mu m.$, $x_0 = \hat{x}_0 = 20\,\mu m.$, $n_e = 2.2$, $\varepsilon_r = 8000$ and $V_A = 1000$ V. Also, we take, $I_b = \hat{I}_b$ and $E_A = \hat{E}_A = 1000$ V/cm.

For studying the coupling effects due to intensity amongst the individual solitons in the crystals P and $\hat{P}$, we need to investigate how the spatial profile of both solitons change due to a modification in intensity of any one of them. Hence, we shall consider three different cases, $\rho = \hat{\rho}$, $\rho > \hat{\rho}$, $\rho < \hat{\rho}$ as shown in Table 6.1 and with these parameters plot the intensity profile for each soliton. The normalized intensities of

Table 6.1 Calculations pertaining to different intensities in crystal P and $\hat{P}$ [3]

	ρ	$\hat{\rho}$	g	$\hat{g}$	β	$\hat{\beta}$
Curve 1	1	1	0.5	0.5	−1.11	−1.11
Curve 2	5	1	0.25	0.75	−0.28	−2.5
Curve 3	1	10	0.85	0.15	−3.18	−0.11

the soliton in crystal P and $\hat{P}$ are shown in Figs. 6.2 and 6.3 respectively. The soliton FWHM rises in crystal P with an increase in its intensity while the soliton FWHM reduces in crystal $\hat{P}$ even if its intensity is kept constant. Similarly, an increase the intensity of the soliton in crystal $\hat{P}$ increases the soliton FWHM in crystal $\hat{P}$ while reducing the FWHM of the soliton in crystal P.

The above analysis is performed when the background intensities are equal in both the crystals, i.e., assuming $I_b = \hat{I}_b$. The effect of the background intensities is non trivial and hence, we shall now consider the case in which the soliton intensities in the two crystals are constant and consider a change in the background intensities. Table 6.2 summarizes the parameters for the aforementioned situation. The normalized intensity for the soliton in both crystals is shown in Figs. 6.4 and 6.5.

Consider a bright soliton propagating in crystal P and a grey soliton propagating in crystal $\hat{P}$ for a grey bright separate coupled soliton pair. In this case, we can obtain the coupling factors,

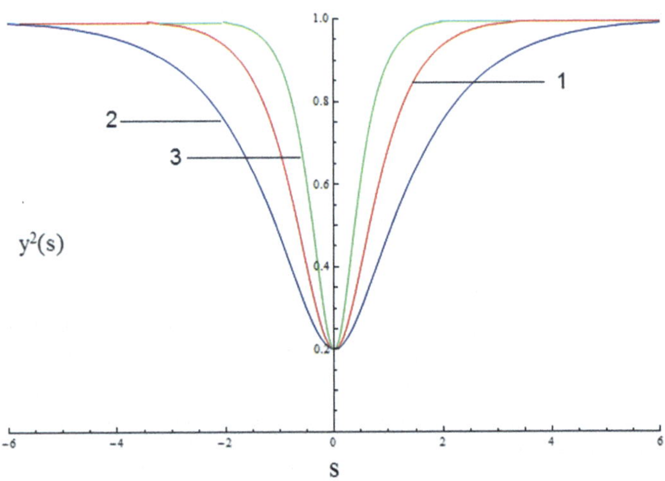

Fig. 6.2 Normalized spatial profiles of the intensity for the grey soliton in P considering various values for the intensities as shown in Table 6.1 (Reprinted by permission from Springer Nature: Springer European Physical Journal D, Coupling effects for separate spatial solitons in a biased series centrosymmetric photorefractive crystal circuit considering grey solitons, Aavishkar Katti and Ram Anjore Yadav, Copyright 2018)

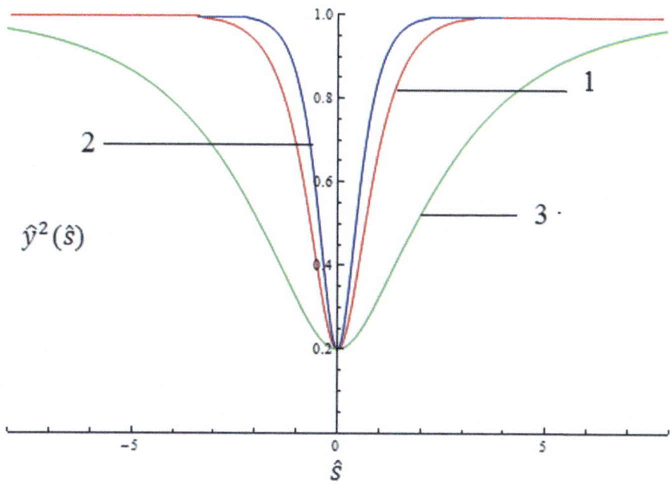

Fig. 6.3 Normalized spatial profiles of the intensity for the grey soliton in $\hat{P}$ considering various values for the intensities as shown in Table 6.1 (Reprinted by permission from Springer Nature: Springer European Physical Journal D, Coupling effects for separate spatial solitons in a biased series centrosymmetric photorefractive crystal circuit considering grey solitons, Aavishkar Katti and Ram Anjore Yadav, Copyright 2018)

Table 6.2 Calculations pertaining to different background intensities in crystal P and $\hat{P}$[3]

	I_b	$\hat{I}_b$	ρ	$\hat{\rho}$	g	$\hat{g}$	β	$\hat{\beta}$
Curve 1	$= I_d = I_\infty/2$	$= \hat{I}_d = \hat{I}_\infty/2$	1	1	0.5	0.5	−1.11	−1.11
Curve 2	$= I_d/2 = I_\infty/4$	$= \hat{I}_d = \hat{I}_\infty/2$	1.33	1	0.53	0.47	−1.25	−0.98
Curve 3	$= I_d = I_\infty/2$	$= \hat{I}_d/10 = \hat{I}_\infty/20$	1	1.82	0.44	0.56	−0.86	−1.40

$$g = \frac{\hat{\delta}\left(\hat{I}_\infty + \hat{I}_b + \hat{I}_d\right)}{\delta(I_b + I_d) + \hat{\delta}\left(\hat{I}_\infty + \hat{I}_b + \hat{I}_d\right)} \tag{6.57}$$

$$\hat{g} = \frac{\delta(I_b + I_d)}{\delta(I_b + I_d) + \hat{\delta}\left(\hat{I}_\infty + \hat{I}_b + \hat{I}_d\right)} \tag{6.58}$$

From (6.57) and (6.58), we can see that the coupling factor g and $\hat{g}$ are functions of the intensity in crystal $\hat{P}$ only. Since we consider a bright soliton propagating in crystal P and a grey soliton propagating in crystal $\hat{P}$, any change in the characteristics of the grey soliton can profoundly affect the characteristics of bright soliton but the reverse is not true, i.e., the grey soliton in $\hat{P}$ cannot be affected by any change in the characteristics of the bright soliton in P.

Now again, we take, $I_b = \hat{I}_b$. For studying the coupling effects due to intensity, we need to see how the change in intensity of the soliton in one crystal affects the

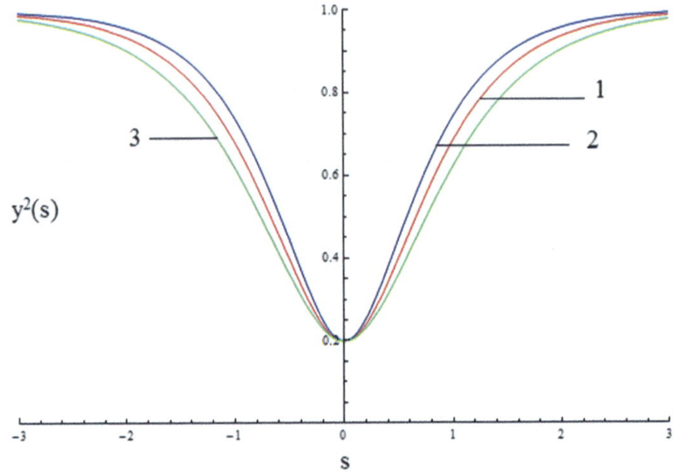

Fig. 6.4 Normalized spatial profiles of the intensity for the grey soliton in p considering various values for the background intensities as shown in Table 6.2 (Reprinted by permission from Springer Nature: Springer European Physical Journal D, Coupling effects for separate spatial solitons in a biased series centrosymmetric photorefractive crystal circuit considering grey solitons, Aavishkar Katti and Ram Anjore Yadav, Copyright 2018)

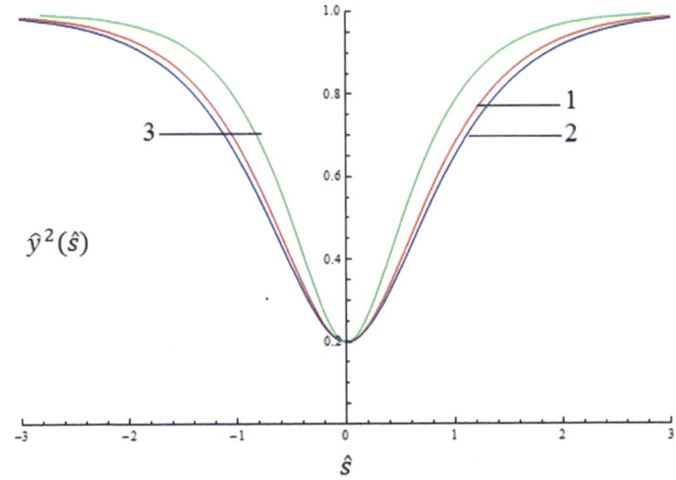

Fig. 6.5 Normalized spatial profiles of the intensity for the grey soliton in $\hat{P}$ considering various values for the background intensities as shown in Table 6.2 (Reprinted by permission from Springer Nature: Springer European Physical Journal D, Coupling effects for separate spatial solitons in a biased series centrosymmetric photorefractive crystal circuit considering grey solitons, Aavishkar Katti and Ram Anjore Yadav, Copyright 2018)

soliton profile in the other crystal. Hence, we shall consider three different cases, $r = \hat{\rho}$, $r > \hat{\rho}$, $r < \hat{\rho}$ as shown in Table 6.3 and with these parameters plot the intensity profile for each soliton. The normalized intensities of the soliton in crystal P and $\hat{P}$ are shown in Figs. 6.6 and 6.7 respectively. We can see from Curve 2 that the grey soliton affects the bright soliton but the bright soliton cannot affect the grey soliton.

Lastly, we shall study the grey dark separate coupled soliton pair. Again, consider $I_b = \hat{I}_b$. For studying the coupling effects due to intensity of the individual soliton in each crystal, we shall consider three different cases, $\rho = \hat{\rho}$, $\rho > \hat{\rho}$, $\rho < \hat{\rho}$ as shown in Table 6.4 and with these parameters plot the intensity profile for each soliton. The normalized spatial profiles for the intensity of the soliton in crystal P and $\hat{P}$ are shown in Figs. 6.8 and 6.9 respectively. The effect of the background intensities for the grey-dark and grey-bright separately coupled solitons can be proceeded exactly as shown previously for the grey grey separately coupled soliton case. It will not be shown here, and the reader is referred to [3] for detailed calculations.

Table 6.3 Calculations pertaining to different intensities in crystal P and $\hat{P}$[3]

	r	$\hat{\rho}$	g	$\hat{g}$	β	$\hat{\beta}$
Curve 1	1	1	2/3	1/3	1.97	−0.49
Curve 2	5	1	2/3	1/3	1.97	−0.49
Curve 3	1	10	6/7	1/7	3.27	−0.09

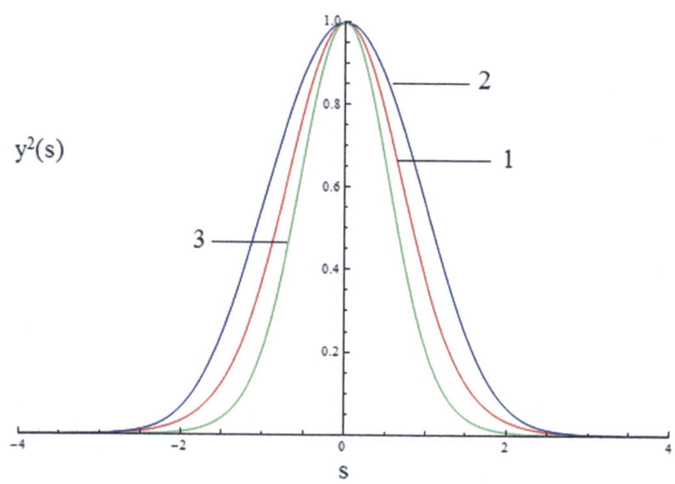

Fig. 6.6 Normalized spatial profiles of the intensity for the bright soliton in P considering various values for the intensities as shown in Table 6.3 (Reprinted by permission from Springer Nature: Springer European Physical Journal D, Coupling effects for separate spatial solitons in a biased series centrosymmetric photorefractive crystal circuit considering grey solitons, Aavishkar Katti and Ram Anjore Yadav, Copyright 2018)

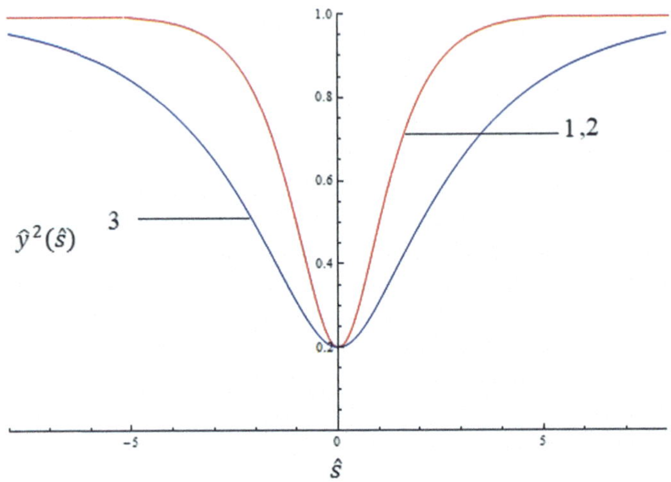

Fig. 6.7 Normalized spatial profiles of the intensity for the grey soliton in $\hat{P}$ considering various values for the intensities as shown in Table 6.3 (Reprinted by permission from Springer Nature: Springer European Physical Journal D, Coupling effects for separate spatial solitons in a biased series centrosymmetric photorefractive crystal circuit considering grey solitons, Aavishkar Katti and Ram Anjore Yadav, Copyright 2018)

Table 6.4 Calculations pertaining to different intensities in crystal P and $\hat{P}$[3]

	ρ	$\hat{\rho}$	g	$\hat{g}$	β	$\hat{\beta}$
Curve 1	1	1	0.5	0.5	−1.11	−1.11
Curve 2	5	1	0.25	0.75	−0.28	−2.5
Curve 3	1	10	0.85	0.15	−3.18	−0.11

6.3.5 Temperature Effects

The temperature of a photorefractive crystal can have a profound effect on self trapping of each soliton. This follows from the fact that the dark irradiance I_d and $\hat{I}_d$ is a function of the temperature of the crystal. Since the photorefractive crystals are connected in series, the temperature of each crystal can affect the self trapping in the other crystal. Cheng and Partovi [8] formulated a model of one level and one carrier to explain how temperature affects the photorefractive mechanism. Any modification in the respective dark irradiances influences the ratio of background intensity to dark irradiance ρ and $\hat{\rho}$. Considering $I_b = \hat{I}_b = 0$, and also $\rho_0 = I_\infty/I_{d0}$ and $\hat{\rho}_0 = \hat{I}_\infty/\hat{I}_{d0}$ which represent ρ and $\hat{\rho}$ at a given temperature of 300 K for both crystals. The parameters ρ and $\hat{\rho}$ are written as [8, 9],

$$\rho = I_\infty/I_d = \rho_0\left(\frac{T}{300}\right)^{-3/2}\exp\left[-\frac{E_t}{k_B}\left(\frac{1}{300}-\frac{1}{T}\right)\right] \tag{6.59}$$

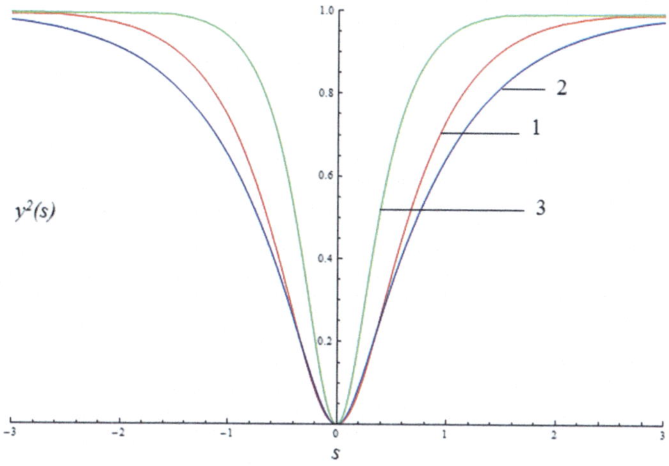

Fig. 6.8 Normalized spatial profiles of the intensity for the grey soliton in P considering various values for the background intensities as shown in Table 6.4 (Reprinted by permission from Springer Nature: Springer European Physical Journal D, Coupling effects for separate spatial solitons in a biased series centrosymmetric photorefractive crystal circuit considering grey solitons, Aavishkar Katti and Ram Anjore Yadav, Copyright 2018)

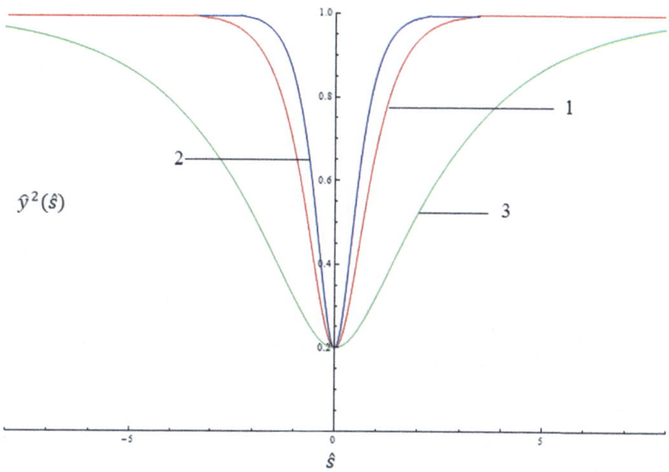

Fig. 6.9 Normalized spatial profiles of the intensity for the grey soliton in $\hat{P}$ considering various values for the background intensities as shown in Table 6.4 (Reprinted by permission from Springer Nature: Springer European Physical Journal D, Coupling effects for separate spatial solitons in a biased series centrosymmetric photorefractive crystal circuit considering grey solitons, Aavishkar Katti and Ram Anjore Yadav, Copyright 2018)

$$\hat{\rho} = \hat{I}_{\infty}/\hat{I}_{d} = \hat{\rho}_0\left(\frac{\hat{T}}{300}\right)^{-3/2}\exp\left[-\frac{E_t}{k_B}\left(\frac{1}{300}-\frac{1}{\hat{T}}\right)\right] \tag{6.60}$$

where $E_{\mathrm{t}} = 10^{-19}$ J is the level location in the energy gap and k_{B} is the Boltzmann constant.

Now, consider a case where T, the temperature of the crystal P is modified while the temperature $\hat{T}$ of the crystal $\hat{P}$ remains unchanged. Using the fact that $\hat{\delta} = \delta$, $I_d = \hat{I}_d$ and $I_b = \hat{I}_b = 0$ and substituting (6.59) and (6.60) into (6.14) and (6.15), we obtain,

$$g = \frac{[1+\hat{\rho}]}{\left[1+\hat{\rho}+1+\rho_0\left(\frac{T}{300}\right)^{-3/2}\exp\left[-\frac{E_t}{k_B}\left(\frac{1}{300}-\frac{1}{T}\right)\right]\right]} \tag{6.61}$$

$$\hat{g} = \frac{1+\rho_0\left(\frac{T}{300}\right)^{-3/2}\exp\left[-\frac{E_t}{k_B}\left(\frac{1}{300}-\frac{1}{T}\right)\right]}{\left[1+\hat{\rho}_0+1+\rho_0\left(\frac{T}{300}\right)^{-3/2}\exp\left[-\frac{E_t}{k_B}\left(\frac{1}{300}-\frac{1}{T}\right)\right]\right]} \tag{6.62}$$

Since the respective coupling factors are functions of the crystal temperature, it is clear that β and $\hat{\beta}$ will also get modified if the temperature of the crystal is modified and in turn alter the self trapping.

Moreover, the dielectric constant is a function of temperature in a paraelectric material. Using the Curie–Weiss law, the dielectric constant can be expressed as the following assuming the mean field approximation is valid [3, 9],

$$\epsilon_r = \epsilon_{r0}\frac{T_0 - T_C}{T - T_C} \tag{6.63}$$

where, $\epsilon_{r0} = \epsilon_r(T = T_0)$.

The typical parameters for KLTN are, $T_{\mathrm{C}} = 279.2$ K, $\epsilon_{r0} = 7.21 \times 103$ at $T_0 = 300$ K [9, 10]. Other parameters of KLTN will be taken as mentioned previously. Since the mean field approximation has to be valid, the working point is selected in such a way that it is far from the phase transition ($Tp = 283$ K). Hence, a temperature range of 290–350 K will be considered for examining the coupling effects of the temperature on the soliton in each crystal. Table 6.5 shows the various calculations as per (6.61)–(6.63) for different temperatures of P and $\hat{P}$.

Table 6.5 Parameters taken for studying the temperature effects on separate coupling [3]

T (Crystal P)	T (Crystal $\hat{P}$)	ρ	$\hat{\rho}$	g	$\hat{g}$
290	300	2.420	1	0.631	0.369
325	300	0.138	1	0.363	0.637
350	300	0.026	1	0.339	0.661

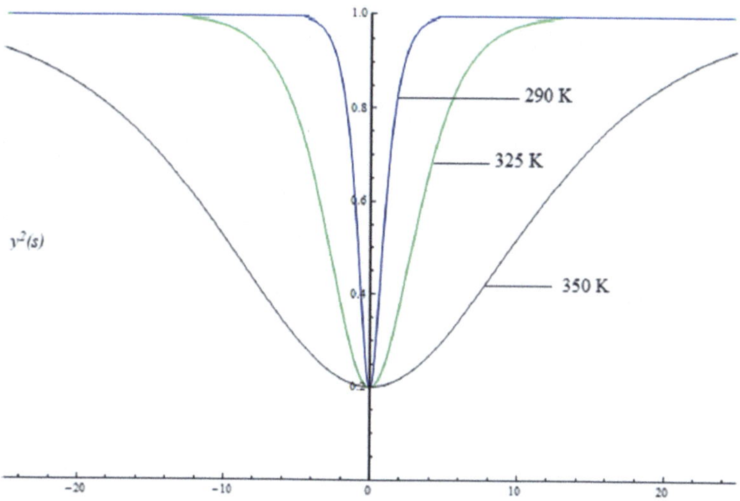

Fig. 6.10 Spatial profiles of the soliton in crystal P for different temperatures of crystal P (Reprinted by permission from Springer Nature: Springer European Physical Journal D, Coupling effects for separate spatial solitons in a biased series centrosymmetric photorefractive crystal circuit considering grey solitons, Aavishkar Katti and Ram Anjore Yadav, Copyright 2018)

Figures 6.10 and 6.11 show the normalized spatial profiles for the intensities of each soliton in P and $\hat{P}$ for the diverse temperature values of crystal P. Each soliton's FWHM is an increasing function of temperature (of P) revealing a weakening of the nonlinearity.

Changing the temperature of $\hat{P}$ and keeping the temperature of P constant will give similar results due to symmetry.

6.3.6 *Some Alternate Configurations*

In this chapter till now, we have illustrated a theory which can explain a wide range of permutations for separate coupled solitons in different configurations of photorefractive crystals. Separately coupled solitons can be bright-bright, dark-dark, bright-dark, grey-grey, grey-bright and grey-dark. The respective ansatz for the bright, dark or grey solitons can be used to solve the paraxial Helmholtz equation in the same manner as illustrated in this chapter. We can consider the photorefractive crystal circuit in which the photorefractive crystals are having the linear electro-optic effect [$a = 1$, $b = 0$ in (6.2)], quadratic electro-optic effect [$a = 0$, $b = 1$ in (6.2)] or exhibiting both simultaneously($a = 1$, $b = 1$). The reader is referred to Refs.[3–7] for detailed investigations in the aforementioned cases. A fundamentally similar but quantitatively different case is that of a photovoltaic crystal circuit which we shall discuss briefly below.

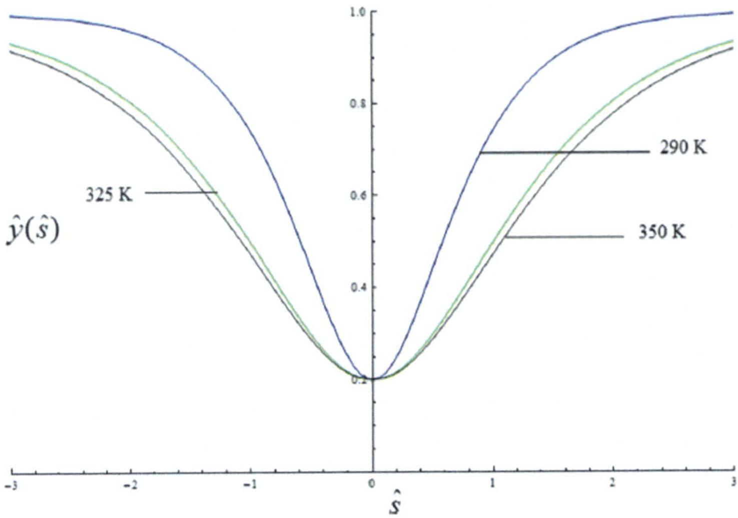

Fig. 6.11 Spatial profiles of the soliton in crystal $\hat{P}$ for different temperatures of crystal P. (Reprinted by permission from Springer Nature: Springer European Physical Journal D, Coupling effects for separate spatial solitons in a biased series centrosymmetric photorefractive crystal circuit considering grey solitons, Aavishkar Katti and Ram Anjore Yadav, Copyright 2018)

6.4 Photovoltaic Crystal Circuits [10]

The photorefractive crystal has a finite photovoltaic coefficient and hence the circuit can be unbiased because the photovoltaic field can provide the current. The charge transport equations are slightly modified [11],

$$\gamma_R n N_D^+ = s_i(I + I_d)\left(N_D - N_D^+\right) \tag{6.64}$$

$$\frac{\partial E_{sc}}{\partial x} = \frac{e}{\epsilon_0 \epsilon}\left(N_D^+ - N_A - n\right) \tag{6.65}$$

$$J = e\mu n E_{sc} + k_B T \mu \frac{\partial n}{\partial x} + k_p s_i\left(N_D - N_D^+\right) I \tag{6.66}$$

$$\frac{\partial J}{\partial x} = 0 \tag{6.67}$$

These are the same as Eqs. (2.2)–(2.5) of Chap. 2. Following the approach of Sect. 2.2.1, we have, for the crystal P,

$$n_\infty = \frac{s_i(N_D - N_A)}{\gamma_R N_A}(I_\infty + I_d) \tag{6.68}$$

$$J_\infty = e\mu n_\infty \left(E_0 + E_p \frac{I_\infty}{I_\infty + I_d} \right) \tag{6.69}$$

with, $E_p = k_p \gamma_R N_A/(e\mu)$.

Similarly for the crystal $\hat{P}$, we have,

$$\hat{n}_\infty = \frac{\hat{s}_i \left(\hat{N}_D - \hat{N}_A \right)}{\hat{\gamma}_R \hat{N}_A} \left(\hat{I}_\infty + \hat{I}_d \right) \tag{6.70}$$

$$\hat{J}_\infty = \hat{e}\hat{\mu}\hat{n}_\infty \left(\hat{E}_0 + \hat{E}_p \frac{\hat{I}_\infty}{\hat{I}_\infty + \hat{I}_d} \right) \tag{6.71}$$

with, $\hat{E}_p = \hat{k}_p \hat{\gamma}_R \hat{N}_A/(\hat{e}\hat{\mu})$.

V, $\hat{V}$ are the voltages measured between the electrodes of the crystals P, $\hat{P}$ respectively having width W, $\hat{W}$. If S, $\hat{S}$ denote the surface area of the electrodes at crystal P, $\hat{P}$ respectively, then $SJ = \hat{S}\hat{J}$ since the crystals are connected in series. Within the approximation of the x-width of the crystals being much greater than the spatial extent of the soliton, $\hat{E}_{sc}(\hat{x} \to \infty) = \hat{E}_0 = \hat{V}/\hat{W}$.and $E_{sc}(x \to \infty) = E_o = V/W$. Again, since the two photorefractive crystals are connected in series and the crystals c-axes are oriented in the right handed screw sense (Fig. 6.12a), $SJ = \hat{S}\hat{J}$ and $V + \hat{V} = 0$. Hence,

$$WE_0 + \hat{W}\hat{E}_0 = 0 \tag{6.72}$$

Again, since the current density is constant in x [from (6.67)]. So, $J_\infty(x \to \infty, z) = J$. Similarly, $\hat{J}_\infty(\hat{x} \to \infty, \hat{z}) = \hat{J}$. For $SJ = \hat{S}\hat{J}$, we have, $SJ_\infty = \hat{S}\hat{J}_\infty$. Substituting (6.69) and (6.71),

$$S\mu n_\infty \left(E_0 + E_p \frac{I_\infty}{I_\infty + I_d} \right) = \hat{S}\hat{\mu}\hat{n}_\infty \left(\hat{E}_0 + \hat{E}_p \frac{\hat{I}_\infty}{\hat{I}_\infty + \hat{I}_d} \right) \tag{6.73}$$

From (6.72) and (6.73), we can easily infer that,

$$E_0 = g\hat{E}_p - \Gamma E_p \tag{6.74}$$

$$\hat{E}_0 = \hat{g}E_p - \hat{\Gamma}\hat{E}_p \tag{6.75}$$

where,

$$g = \frac{\hat{\delta}\hat{I}_\infty \hat{W}}{W\left[\delta(I_\infty + I_d) + \hat{\delta}\left(\hat{I}_\infty + \hat{I}_d\right)\right]}, \Gamma = \frac{\delta I_\infty}{\left[\delta(I_\infty + I_d) + \hat{\delta}\left(\hat{I}_\infty + \hat{I}_d\right)\right]},$$

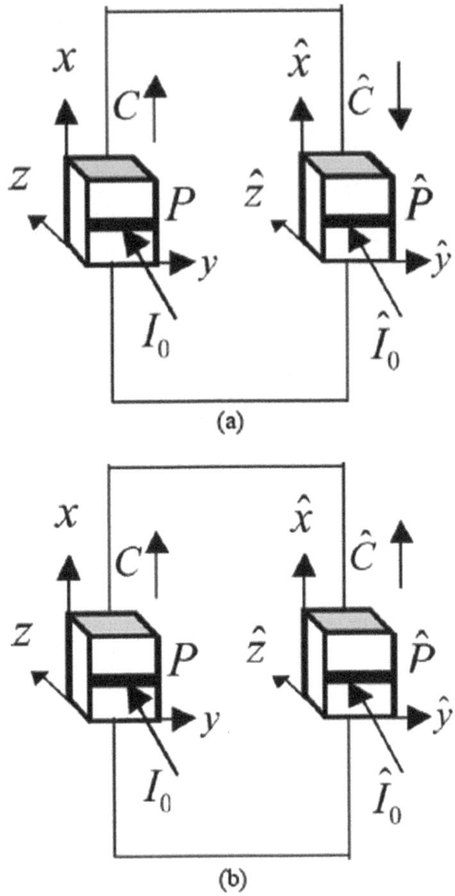

Fig. 6.12 Configuration for unbiased photovoltaic crystal circuit (Reprinted from Physics Letters A, 300, Jinsong Liu, Separate spatial soliton pairs and solitons interaction in an unbiased series photorefractive crystal circuit, 213–220, Copyright 2002, with permission from Elsevier)

$$\hat{g} = \frac{\delta I_\infty W}{\hat{W}\left[\delta(I_\infty + I_d) + \hat{\delta}\left(\hat{I}_\infty + \hat{I}_d\right)\right]}$$

$$\delta = S\mu s_i(N_D - N_A)/(\gamma_R N_A W), \hat{\delta} = \hat{S}\hat{\mu}\hat{s}_i\left(\hat{N}_D - \hat{N}_A\right)/\left(\hat{\gamma}_R \hat{N}_A \hat{W}\right)$$

g and $\hat{g}$ are identified as gain factors while $\Gamma, \widehat{\Gamma}$ are identified as the coupling factors. Now if the c-axes are oriented in the opposite screw senses (Fig. 6.12b), $V + \hat{V} = 0$ and $SJ = -\hat{S}\hat{J}$. This is because an illuminated photovoltaic photorefractive crystal is like a current source where current flows out from its positive electrode. As in the previous case, we have,

$$E_0 = -g\hat{E}_p - \Gamma E_p \tag{6.76}$$

$$\hat{E}_0 = -\hat{g}E_p - \widehat{\Gamma}\hat{E}_p \tag{6.77}$$

When all the parameters of the two crystals are equal for this case, it is notable that the current in the circuit will be zero. From (6.64) and (6.66),

$$J = e\mu n\left(E_{sc} + \frac{k_B T}{e}\frac{\partial \ln n}{\partial x} + E_p \frac{I}{I + I_d}\right) \tag{6.78}$$

Since $J(x, z) = J_\infty$, from (6.69) and (6.78),

$$n_\infty\left(E_p \frac{I_\infty}{I_\infty + I_d} + E_0\right) = n\left(E_{sc} + \frac{k_B T}{e}\frac{\partial \ln n}{\partial x} + E_p \frac{I}{I + I_d}\right) \tag{6.79}$$

If the diffusion effect is neglected,

$$E_{sc} = E_p \frac{I_\infty - I}{I + I_d} + E_0 \frac{I_\infty + I_d}{I + I_d} \tag{6.80}$$

And, for the crystal $\hat{P}$,

$$\hat{E}_{sc} = E_p \frac{\hat{I}_\infty - \hat{I}}{\hat{I} + \hat{I}_d} + \hat{E}_0 \frac{\hat{I}_\infty + \hat{I}_d}{\hat{I} + \hat{I}_d} \tag{6.81}$$

E_{sc} and $\hat{E}_{sc}$ are not independent of each other as we can see from (6.74) and (6.75). As done in [10], we can now use the space charge fields in the paraxial Helmholtz equation to solve for bright, dark and bright-dark separately coupled photovoltaic solitons following a similar approach as elucidated in this chapter before.

6.5 Concluding Remarks and Further Reading

In conclusion, we have delineated a general theory for optical spatial solitons propagating in a biased photorefractive crystal circuit. We have then extended and investigated the validity of the theory for various different types and configurations of the constituent photorefractive crystals. The coupled space charge fields in both the crystals of the photorefractive crystal circuit have been derived. The input intensity of each soliton and the temperature of each crystal exert a coupling effect between the individual solitons in the two crystals. These phenomena are discussed in detail taking relevant examples.

Further investigations of separately coupled solitons include the study of dynamical evolution and the stability of these solitons. For studying the dynamical evolution, we shall first consider the field profiles as found in (6.23), i.e., $y_0(s)$ as found in (6.30) for the soliton in crystal P and $\hat{y}_0(s)$ from (6.32) for the soliton in crystal $\hat{P}$. The grey solitary states will be, $U_0 = \sqrt{1}y_0(s)\exp\left[i\left(\int_0^s \frac{Qd\bar{s}}{y^2(\bar{s})}\right)\right]$ and $\hat{U}_0 = \sqrt{1}\hat{y}_0(\hat{s})\exp\left[i\left(\int_0^{\hat{s}} \frac{Qd\bar{s}}{y^2(\bar{s})}\right)\right]$ with $\rho = \hat{\rho} = 1$, $\beta = \hat{\beta} = -1.11$. Consider these two solitary states as input beams. The dynamical evolution of both solitons can be investigated by solving Eqs. (6.20) and (6.21) by using a finite difference beam propagation method or a split step fourier method as has been performed in [5].

For considering the stabilities, we can consider a similar beam profile but with a increased amplitude, i.e., $U_1 = \sqrt{10}y_0(s)\exp\left[i\left(\int_0^s \frac{Qd\bar{s}}{y^2(\bar{s})}\right)\right]$. We need to see the dynamical evolution of this beam U_1 in crystal P and $U_0 = \sqrt{1}y_0(s)\exp\left[i\left(\int_0^s \frac{Qd\bar{s}}{y^2(\bar{s})}\right)\right]$ in crystal $\hat{P}$ simultaneously by taking these as input beams and applying a beam propagation method in (6.20) and (6.21). Since U_I is not a solitary state of (6.30), we have included a perturbation from which we can investigate whether it grows or decays while propagating [5].

With regards to the different configurations of the photorefractive crystal circuits in which separate coupled solitons have been studied, the reader is referred to Refs. [4, 7], where the authors investigate separate coupling in novel photorefractive crystal circuit consisting of both linear and quadratic nonlinearity, Ref. [10] in which the author studies photovoltaic crystal circuits in much more detail, also Refs. [6, 12, 13].

References

1. D.N. Christodoulides, S.R. Singh, M.I. Carvalho, M. Segev, Incoherently coupled soliton pairs in biased photorefractive crystals. Appl. Phys. Lett. **68**(13), 1763–1765 (1996). https://doi.org/10.1063/1.116659
2. Z. Chen, M. Segev, T.H. Coskun, D.N. Christodoulides, Y.S. Kivshar, V.V. Afanasjev, Incoherently coupled dark-bright photorefractive solitons. Opt. Lett. **21**(22), 1821–1823 (1996). https://doi.org/10.1364/OL.21.001821
3. A. Katti, R.A. Yadav, Coupling effects for separate spatial solitons in a biased series centrosymmetric photorefractive crystal circuit considering grey solitons. Eur. Phys. J. D **72**(2), 37 (2018). https://doi.org/10.1140/epjd/e2017-80396-x
4. A. Katti, R.A. Yadav, Coupling effects for grey separate spatial solitons in a biased series photorefractive crystal circuit with both the linear and quadratic electro-optic effects. Opt. Quant. Electron. **49**(1), 36 (2017). https://doi.org/10.1007/s11082-016-0887-1
5. J.S. Liu, Z.H. Hao, Evolution of separate screening soliton pairs in a biased series photorefractive crystal circuit. Phys. Rev. E **65**(6), 066601 (2002). https://doi.org/10.1103/PhysRevE.65.066601
6. K. Zhan, C. Hou, Separate spatial soliton pairs in a biased series centrosymmetric photorefractive crystal circuit. Optik **122**(7), 563–568 (2011)
7. Q. Jiang, Y. Su, H. Nie, Z. Ma, Y. Li, Separate spatial soliton pairs in a biased series photorefractive crystal circuit with both the linear and quadratic electro-optic effects. J. Mod. Opt. **64**(6), 609–615 (2017). https://doi.org/10.1080/09500340.2016.1253881
8. L. Cheng, A. Partovi, Temperature and intensity dependence of photorefractive effect in GaAs. Appl. Phys. Lett. **49**(21), 1456–1458 (1986). https://doi.org/10.1063/1.97301

9. K. Zhan, C. Hou, H. Tian, Effects of the temperature on steady-state bright spatial solitons in biased centrosymmetric photorefractive crystals. Appl. Phys. B: Lasers Opt. **100**(4), 821–826 (2010). https://doi.org/10.1007/s00340-010-4067-x
10. A. Katti, Incoherently coupled Gaussian soliton pairs in biased photorefractive crystal having both the linear and quadratic electro-optic effect. Appl. Phys. B **124**(9), 192 (2018)
11. J. Liu, Z. Hao, Soliton parametric coupling in a series photorefractive crystal circuit. Phys. Lett. Sect. A: Gen. Atomic Solid State Phys. **309**(1–2), 44–52 (2003). https://doi.org/10.1016/S0375-9601(03)00172-5
12. L. Keqing, T. Tiantong, Z. Yanpeng, One-dimensional steady-state spatial solitons in photovoltaic photorefractive materials with an external applied field. Phys. Rev. A **61**(5), 053822 (2000). https://doi.org/10.1103/PhysRevA.61.053822
13. Z. Hao, J. Liu, Interaction between separate spatial soliton pairs in a biased serial photovoltaic photorefractive crystal circuit and its application, in *Photonics Asia 2002* (2002), pp. 326–335. Available https://proceedings.spiedigitallibrary.org/proceeding.aspx?articleid=885450

Chapter 7
Photorefractive Waveguides

7.1 Introduction

We have seen previously that a spatial soliton traversing in a photorefractive crystal induces a waveguide which then guides the soliton beam itself [1, 2]. An interesting case arises if we embed a planar waveguide inside a photorefractive crystal. The diffraction effect will be counteracted due to the waveguiding of the planar waveguide. So, in a photorefractive waveguide, the minimum power required for soliton formation or the threshold power is much lower as compared to that in conventional photorefractive media. There has been a recent upsurge in interest in such photorefractive waveguides because of the potential important implications for practical applications [3–7]. Optical spatial solitons have important uses in various applications such as optical switching, routing, waveguiding and navigations etc. Also, in contrast to temporal solitons, the threshold power required is quite large for spatial solitons. This difficulty can be overcome simply if the spatial solitons are created in a waveguide. The self defocusing is partly eliminated due to the self focusing effect of the waveguide and the formation of optical spatial solitons takes place with relatively lower powers.

In this chapter, we shall study the propagation characteristics of bright spatial solitons in a biased planar photorefractive waveguide having both the linear and quadratic electro-optic effect under the WKBJ approximation. We shall use the paraxial wave equation incorporating the waveguiding effect through a waveguide parameter. Then we shall assume a variational solution of a quasi-soliton and proceed towards determining the variational parameters. Finally, we assign a physical interpretation to each parameter and analyze the quasi-soliton characteristics and the effect of the waveguide parameter on the self trapping.

A. Katti and R. A. Yadav, *Optical Spatial Solitons in Photorefractive Materials*, Progress in Optical Science and Photonics 14, https://doi.org/10.1007/978-981-16-2550-3_7

7.2 Mathematical Formulation

We shall consider an optical beam propagating along the z direction in a photorefractive crystal. In addition, a waveguide has been embedded in the photorefractive crystal. The photorefractive crystal exhibits the linear electro-optic effect, the quadratic electro-optic effect or both the linear and quadratic electro-optic effect simultaneously. The optical c-axis of the photorefractive waveguide is considered to be along the x-axis. The beam is polarized along the x-direction we consider the diffraction along the same axis. The external bias is along the x-direction and hence a space charge field $\overrightarrow{E}_{sc} = \hat{x}E_{sc}$ is set up in the photorefractive waveguide.

The electric field $\overrightarrow{E}$ of the incident beam propagating through the photorefractive waveguide satisfies the following wave equation [5, 6],

$$\nabla^2 \overrightarrow{E} + (k_0 n_e')^2 \overrightarrow{E} - gx^2 \overrightarrow{E} = 0 \tag{7.1}$$

where n_e' is the perturbed extraordinary index of refraction,k_0 is the wave number in free space. Since, $\Delta n = n_e' - n_e = -\frac{1}{2}an_e^3 r_{eff} E_{sc} - \frac{1}{2}bn_e^3 g_{eff} \varepsilon_0^2 (\varepsilon_r - 1)^2 E_{sc}^2$, n_e' can be expressed in first order approximation as [8],

$$n_e'^2 = n_e^2 - an_e^4 r_{eff} E_{sc} - bn_e^4 g_{eff} \in_0^2 (\in_r - 1)^2 E_{sc}^2 \tag{7.2}$$

r_{eff} and g_{eff} are the linear and quadratic electro-optic coefficients respectively. a and b will be non zero depending on the presence of the linear or quadratic electro-optic effect respectively. g is a real and positive parameter which we call as the waveguide parameter. This waveguide parameter represents the strength of the waveguiding which counteracts the diffraction of the light beam. In (7.1), the third term which is a function of g is a permanent change of the extraordinary refractive index due to the embedded waveguide. The incident beam envelope is,

$$\overrightarrow{E}(x, z) = \hat{x}\Phi(x, z)e^{ikz} \tag{7.3}$$

where $\Phi(x, z)$ is the slowly varying envelope of the wave and $k = k_0 n_e'$. Applying the paraxial approximation and putting (7.2) and (7.3) in (7.1), (see Note 1 at the end of the chapter),

$$i\frac{\partial \Phi}{\partial z} + \frac{1}{2k_0 n_e}\frac{\partial^2 \Phi}{\partial x^2} - \frac{1}{2}k_0 n_e^3 r_{eff} E_{sc}\Phi - \frac{1}{2}k_0 n_e^3 g_{eff} \in_0^2 (\in_r - 1)^2 E_{sc}^2 \Phi - gx^2\Phi = 0 \tag{7.4}$$

The space charge field disregarding the diffusion effect is [9],

$$E_{sc} = \frac{I_\infty + I_d}{I + I_d} E_0 \tag{7.5}$$

where the symbols have their usual meaning as defined previously.

Using the usual dimensionless coordinates specified before, the space charge field becomes,

$$E_{sc} = \frac{1+\rho}{1+|U|^2} E_0 \tag{7.6}$$

and the evolution equation is obtained,

$$i\frac{\partial U}{\partial \xi} + \frac{1}{2}\frac{\partial^2 U}{\partial s^2} - \beta_1 \frac{1+\rho}{1+|U|^2} U - \beta_2 \left(\frac{1+\rho}{1+|U|^2}\right)^2 U - \delta s^2 U = 0 \tag{7.7}$$

where,

$\beta_1 = a\frac{(k_0 x_0)^2 n_e^4 r_{eff}}{2} E_0$, $\beta_2 = b\frac{(k_0 x_0)^2 n_e^4 g_{eff} \epsilon_0^2 (\epsilon_r - 1)^2}{2} E_0^2$ and $\delta = g k_0 x_0^4 n_e$ along with $\rho = I_\infty / I_d$. The origin of the refractive index perturbation lies in the nonlinear terms in (7.7). Equation (7.7) does not have an exact solution so we have to resort to approximate methods. There are several methods like Akhmanov's paraxial method [10], Segev's method [11], Anderson's variational method [12] and Vlasov's moment method [9] which can be used to solve (7.7). We use a variational solution along with a paraxial approximation to obtain soliton solutions which are acceptable physically. Also $\rho = 0$ since bright solitons will be considered. The slowly varying beam envelope to be expressed as,

$$U(\xi, s) = U_0(\xi, s) e^{-i\Omega(\xi, s)} \tag{7.8}$$

$U_0(\xi, s)$ is a real quantity and $\Omega(\xi, s)$ gives the phase. Substituting (7.8) in (7.7) we obtain,

$$\left(i\frac{\partial U_0}{\partial \xi} + U_0 \frac{\partial \Omega}{\partial \xi}\right) + \frac{1}{2}\left\{\frac{\partial^2 U_0}{\partial s^2} - 2i\frac{\partial U_0}{\partial s}\frac{\partial \Omega}{\partial s} - iU_0\frac{\partial^2 \Omega}{\partial s^2} - U_0\left(\frac{\partial \Omega}{\partial s}\right)^2\right\}$$
$$- \beta_1 \frac{1+\rho}{\left(1+U_0^2\right)} U_0 - \beta_2 \left(\frac{1+\rho}{1+U_0^2}\right)^2 U_0 - \delta s^2 U_0 = 0 \tag{7.9}$$

(7.9) is an equation which is a combination of real and imaginary terms. For the LHS to be zero, we need to equate the real and imaginary parts separately to zero,

$$\frac{\partial U_0}{\partial \xi} - \frac{\partial U_0}{\partial s}\frac{\partial \Omega}{\partial s} - \frac{1}{2}U_0 \frac{\partial^2 \Omega}{\partial s^2} = 0 \tag{7.10}$$

$$U_0 \frac{\partial \Omega}{\partial \xi} + \frac{1}{2}\frac{\partial^2 U}{\partial s^2} - \frac{1}{2}U_0\left(\frac{\partial \Omega}{\partial s}\right)^2 - \beta_1 \Phi_1(\xi, s) U_0 - \beta_2 \Phi_2(\xi, s) U_0 - \delta s^2 U_0 = 0 \tag{7.11}$$

where,

$$\Phi_1(\xi, s) = \frac{1}{1 + |U_0|^2} \tag{7.12}$$

$$\Phi_2(\xi, s) = \frac{1}{(1 + |U_0|^2)^2} \tag{7.13}$$

$\Phi_1(\xi, s)$ and $\Phi_2(\xi, s)$ represent the contribution of the linear and quadratic electro-optic effect to the refractive index change. As mentioned previously, the last term in Eq. (7.11) represents the effect of the embedded planar waveguide. The interplay between the refractive index waveguide and the planar waveguide structure can be seen clearly from the last three terms in (7.11).

7.3 Spatial Solitons

We need to search for physically acceptable bright soliton states, which implies an intensity profile which peaks at the center of the beam and falls off with distance. Following the approach in Chap. 5 (Sect. 5.5), we shall assume a quasi soliton solution, i.e., a modified Gaussian ansatz for the soliton envelope in (7.11) as follows,

$$U_0(\xi, s) = \frac{U_{00}}{\sqrt{f(\xi)}} e^{-s^2/2r^2 f^2(\xi)} \tag{7.14}$$

$$\Omega(\xi, s) = \frac{s^2}{2}\Gamma(\xi) \tag{7.15}$$

$$\Gamma(\xi) = -\frac{1}{f(\xi)}\frac{df(\xi)}{d\xi} \tag{7.16}$$

where $P_0 = U_{00}^2$ is the normalized peak power of the soliton, r is a constant which is always positive, $f(\xi)$ is the variable parameter proportional to the beam width. The spatial width of the soliton is expressed as the product $rf(\xi)$. We term the solution (7.14) as a variational solution where the variable parameters to be found are r, $f(\xi)$, $\Gamma(\xi)$. In general, we shall assume that $\frac{df}{d\xi} = 0$ at $\xi = 0$, i.e., the soliton beam is non-diverging at the entry point of the crystal. Also, we can assume that $f = 1$ at $\xi = 0$. The next step is to simplify the non-linear terms Φ_1 and Φ_2. Expanding both in a Taylor series, we get, (see Note 2 at the end of chapter)

$$\Phi_1(\xi, s) \approx \frac{1}{\left(1 + \frac{U_{00}^2}{f}\right)} + s^2 \frac{\frac{U_{00}^2}{r^2 f^3(\xi)}}{\left(1 + \frac{U_{00}^2}{f(\xi)}\right)^2} \tag{7.17}$$

$$\Phi_2(\xi, s) \approx \frac{1}{\left(1+\frac{U_{00}^2}{f}\right)^2} + 2s^2 \frac{\frac{U_{00}^2}{r^2 f^3(\xi)}}{\left(1+\frac{U_{00}^2}{f(\xi)}\right)^3} \tag{7.18}$$

where we approximate to the first order.

Substituting (7.14)–(7.18) in (7.11), we obtain an equation in various powers of s^2. Considering the coefficients of s^2 in LHS and RHS and equating them, we get,

$$\frac{d^2 f(\xi)}{d\xi^2} = \frac{1}{r^4 f^3(\xi)} - 2\beta_1 \frac{\frac{P_0}{r^2 f^2(\xi)}}{\left(1+\frac{P_0}{f(\xi)}\right)^2} - 4\beta_2 \frac{\frac{P_0}{r^2 f^2(\xi)}}{\left(1+\frac{P_0}{f(\xi)}\right)^3} - 2\delta f(\xi) \tag{7.19}$$

(7.19) can said to be an equation which details the spatial evolution of the parameter $f(\xi)$. Higher order terms of s, i.e., s^3, s^4 will not be considered since we take the first order approximation **$[r f(\xi)] > s$** in (7.17)–(7.18). Now, the light beam's propagation in the photorefractive crystal may proceed in the following three ways: it may travel stably with an unchanging intensity profile, it may diverge, or it may be compressed. From (7.19), we can see that this depends upon the magnitudes of the power P_0 and the parameters β_1, β_2. A soliton, or a stable self trapped beam needs the beam width parameter $f(\xi)$ to remains constant with propagation. Hence equating the LHS in (7.19) to zero,

$$\frac{1}{r^4} = 2\delta + \frac{2\beta_1 P_{0t}}{r^2(1+P_{0t})^2} + \frac{4\beta_2 P_{0t}}{r^2(1+P_{0t})^3} \tag{7.20}$$

In (7.20), we see an equilibrium condition which is known as quartic equilibrium condition because it is a polynomial equation of fourth order and has four roots. From (7.20), we can find out the threshold power P_{0t} needed for stationary propagation of the optical beam. The region of existence of optical spatial solitons in the photorefractive waveguide can be found from (7.20) and hence it is known as the existence equation. Examining the solutions of the above equation, two roots are imaginary, one root is negative and only one root is positive. Neglecting the imaginary and negative values of r for spatial solitons as being unphysical, we are left with only one solution.

7.3.1 *Waveguides in Photorefractive Crystals Having Both Electro-Optic Effects*

For illustration of soliton behavior in such waveguides, PMN-0.33PT crystal will be considered which exhibits both electro-optic effects simultaneously. The parameters taken for the aforementioned crystal are shown in Table 7.1. Figure 7.1 shows the

Table 7.1 Parameters taken in the our calculation [5][a]

n_e	2.562	l (crystal width)	1 cm
x_0	20 μm	$V(bias)$	2000 V
λ_0	632.8 nm	ρ	0
r_{eff}	182×10^{-12} m/V	β_1	30.9224
$g_{eff}\,\epsilon_0^2\,(\epsilon_r - 1)^2$	$1.38 \times 10^{-16}\,m^2/V^2$	β_2	4.6896

[a]Reprinted from Katti [5], Copyright 2018, with permission from Elsevier

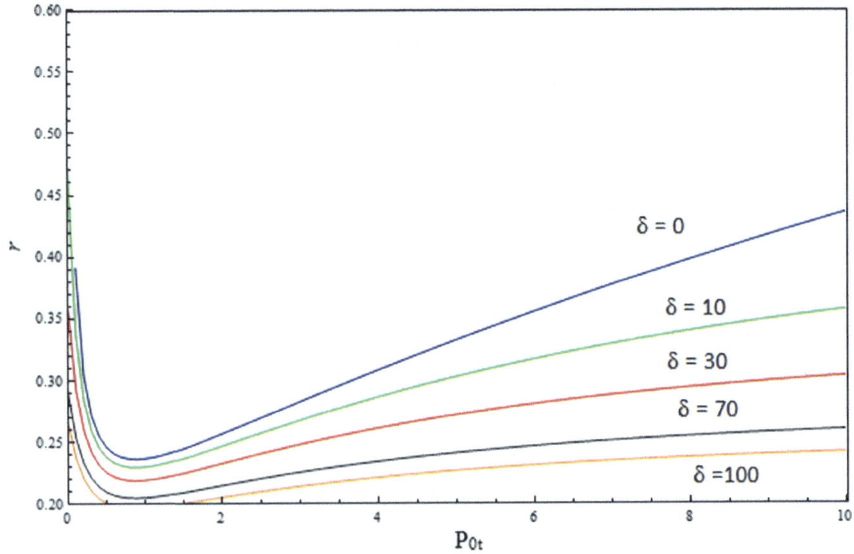

Fig. 7.1 The dependence of the r the stable spatial width of the soliton on the threshold power of the spatial solitons,$\beta_1 = 30.9224$, $\beta_2 = 4.6896$. (Reprinted from Wave Motion, 77, Aavishkar Katti, R.A. Yadav, Awadhesh Prasad, Bright optical spatial solitons in photorefractive waveguides having both the linear and quadratic electro-optic effect, 64–76, Copyright 2018, with permission from Elsevier)

graph of r versus the threshold power P_{0t} for various strengths of the waveguide. The spatial width of the soliton first decreases with an increase in power when the power is low, while the soliton width increases with an increase in power when the power is high. A similar dependence remains for each value of the waveguide parameter but the curve becomes less steep as the waveguide parameter increases. The presence of bistable states is clear from Fig. 7.1 since we can infer two values for the threshold powers for a single value of r and $f(\xi)$ is constant. The two values of the threshold power at which a soliton can just form are known as P_{0t1} and P_{0t2}.

In the absence of waveguiding, we shall see the behavior of solitons in distinct power regions. Consider four values of power, $P_1(= 0.0999) <$

$P_{0t1}, P_2(= 0.2692) = P_{0t1}, P_{0t1} < P_3(= 1) < P_{0t2}, P_4(= 26.92) > P_{0t2}$. In Fig. 7.2, the change in f (which is a parameter related to the soliton width) with the scaled propagation distance ξ is plotted in the four power regimes. Figure 7.2 shows that the beam width deviates to a large value with propagation when $P = P_1$. Since P_1 is less than the threshold power, so we do not expect the soliton to form. If the power exactly equals the threshold power, $= P_{0t1}$, the parameter $f = 1$ and hence the soliton width rf is constant in Fig. 7.2. So we obtain a spatial soliton which travels without changing its shape, i.e., a stable self trapping. If the value of the peak power P lies between the two threshold powers, i.e., $P_{0t2} < P < P_{0t1}$, we find from Fig. 7.2 that the soliton's width oscillates. This oscillation is with amplitude less than unity so we can conclude that there is a reasonable amount of self trapping and this is in fact a "soliton". If the soliton peak power P is larger than P_{0t2}, the soliton width again oscillates with propagation but now the oscillation amplitude is greater than unity and hence it cannot be termed as a soliton. As rf is the soliton width, we use it to plot the soliton's propagation in Figs. 7.3, 7.4, 7.5 and 7.6 using (7.14).

In order to investigate what effect waveguiding has on the self trapping, we shall now take a finite nonzero value for the waveguide parameter and investigate the propagation of the soliton. How the soliton width parameter f changes while propagating is plotted in Fig. 7.3 considering various values of the waveguide parameters δ. If we take, $\delta = 0$ implying no waveguide effect at all, the wave diverges as seen in Fig. 7.3. Again, as the value δ increases, we can see in Fig. 7.3 that the behavior of f changes. For a certain value of δ, the value of f is seen to be oscillatory with respect

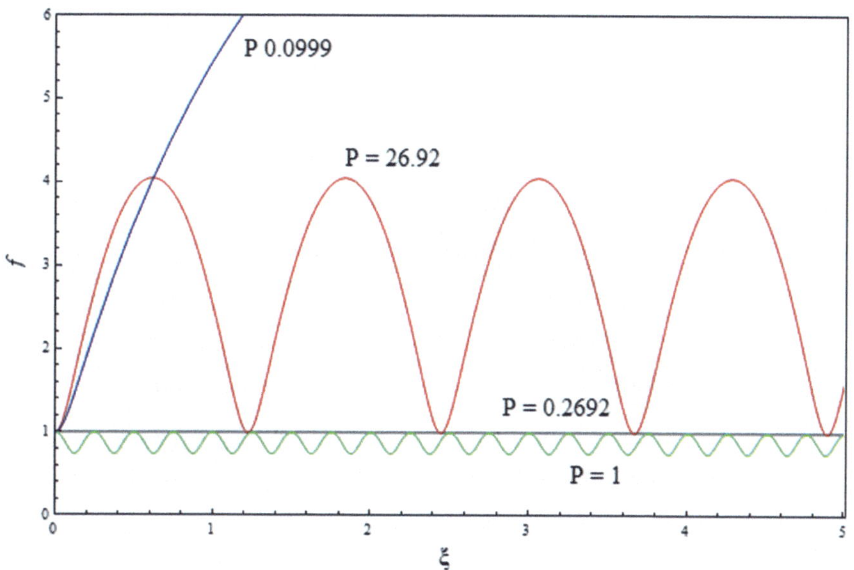

Fig. 7.2 Variable beam width parameter $f(\xi)$ versus the normalized distance of propagation ξ considering various peak powers of the soliton, $\beta_1 = 30.9224$, $\beta_2 = 4.6896$, $r = 0.2793$. (Reprinted from Katti [5], Copyright 2018, with permission from Elsevier)

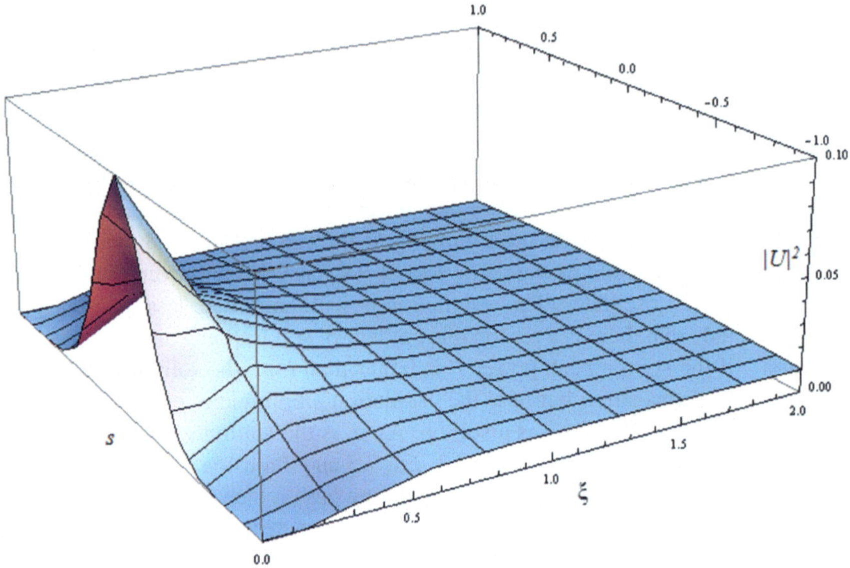

Fig. 7.3 Propagation of the soliton when there is no embedded waveguide. P = 0.0999. $\beta_1 = 30.9224$, $\beta_2 = 4.6896$, $r = 0.2793$. (Reprinted from Wave Motion, 77, Aavishkar Katti, R.A. Yadav, Awadhesh Prasad, Bright optical spatial solitons in photorefractive waveguides having both the linear and quadratic electro-optic effect, 64–76, Copyright 2018, with permission from Elsevier)

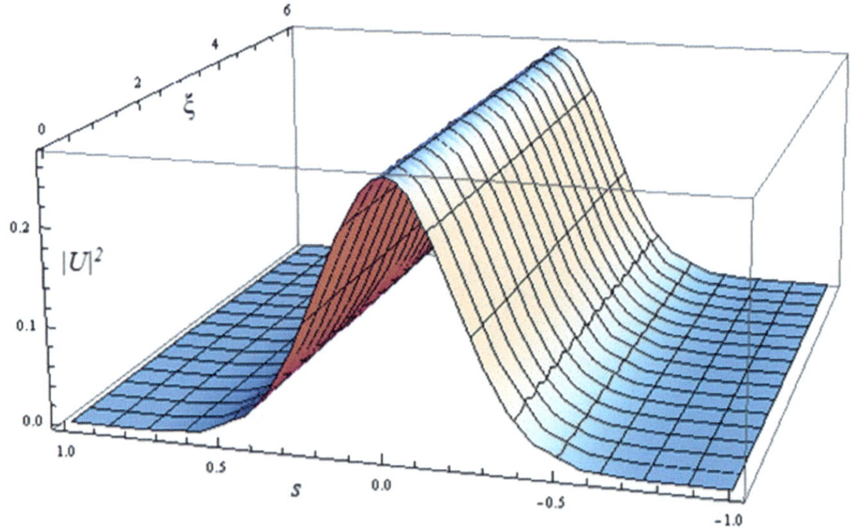

Fig. 7.4 Propagation of the soliton when there is no embedded waveguide. $P = 0.2692$. $\beta_1 = 30.9224$, $\beta_2 = 4.6896$, $r = 0.2793$. (Reprinted from Wave Motion, 77, Aavishkar Katti, R.A. Yadav, Awadhesh Prasad, Bright optical spatial solitons in photorefractive waveguides having both the linear and quadratic electro-optic effect, 64–76, Copyright 2018, with permission from Elsevier)

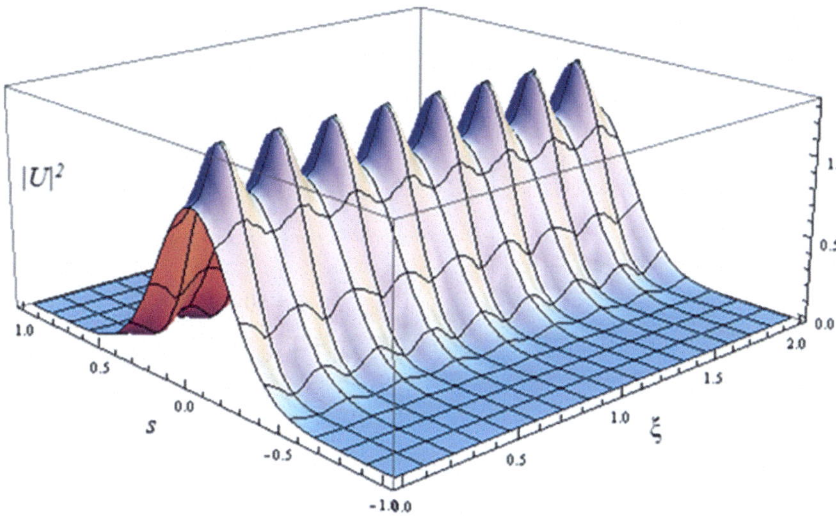

Fig. 7.5 Propagation of the soliton when there is no embedded waveguide. $P = 1$. $\beta_1 = 30.9224$, $\beta_2 = 4.6896$, $r = 0.2793$. (Reprinted from Katti [5], Copyright 2018, with permission from Elsevier)

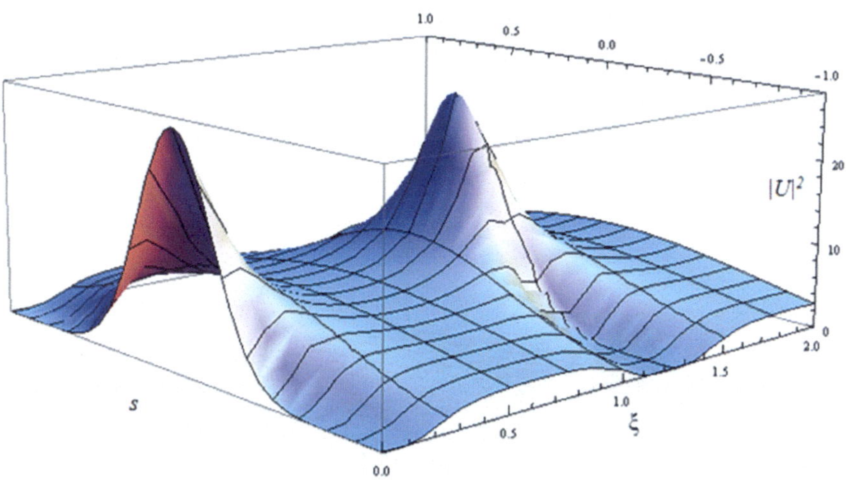

Fig. 7.6 Propagation of the soliton when there is no embedded waveguide. $P = 26.92$. $\beta_1 = 30.9224$, $\beta_2 = 4.6896$, $r = 0.2793$. (Reprinted from Wave Motion, 77, Aavishkar Katti, R.A. Yadav, Awadhesh Prasad, Bright optical spatial solitons in photorefractive waveguides having both the linear and quadratic electro-optic effect, 64–76, Copyright 2018, with permission from Elsevier)

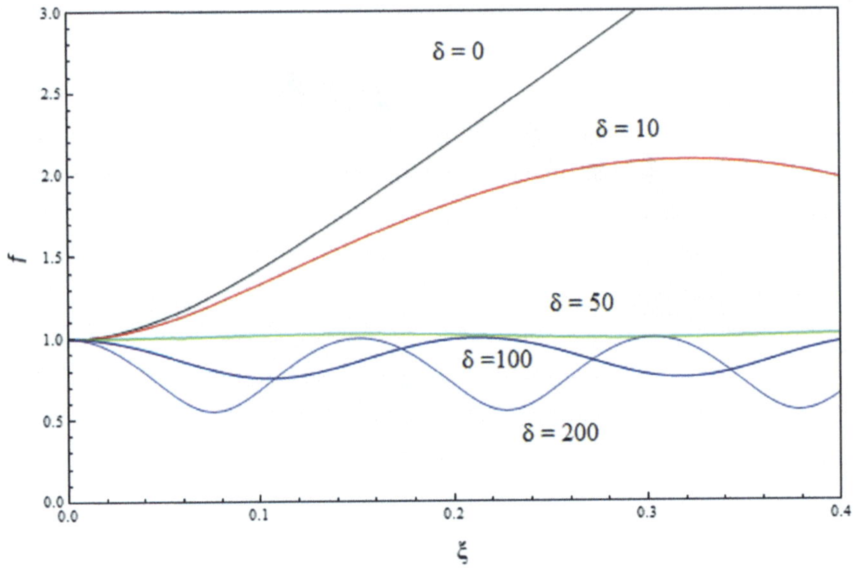

Fig. 7.7 Beam width parameter $f(\xi)$ versus propagation distance ξ for different waveguide strengths, $\beta_1 = 30.9224$, $\beta_2 = 4.6896$, $P_0 = 0.055$ and $r = 0.2831$. (Reprinted from Wave Motion, 77, Aavishkar Katti, R.A. Yadav, Awadhesh Prasad, Bright optical spatial solitons in photorefractive waveguides having both the linear and quadratic electro-optic effect, 64–76, Copyright 2018, with permission from Elsevier)

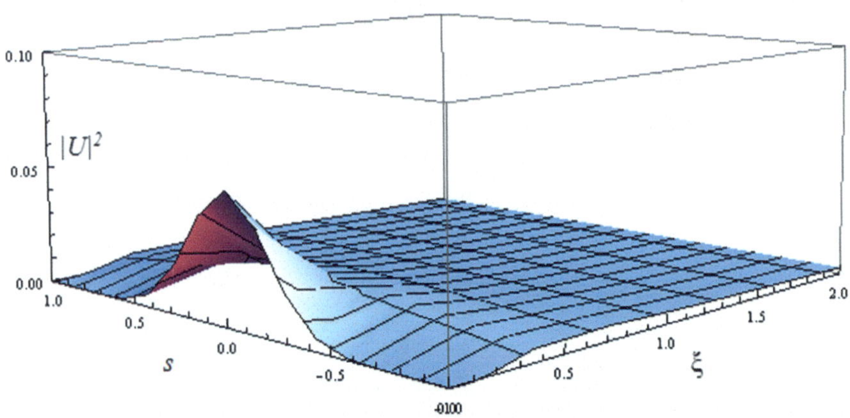

Fig. 7.8 Propagation of soliton in the absence of the waveguide when P = 0.055 and $r = 0.2831$. (Reprinted from Wave Motion, 77, Aavishkar Katti, R.A. Yadav, Awadhesh Prasad, Bright optical spatial solitons in photorefractive waveguides having both the linear and quadratic electro-optic effect, 64-76, Copyright 2018, with permission from Elsevier)

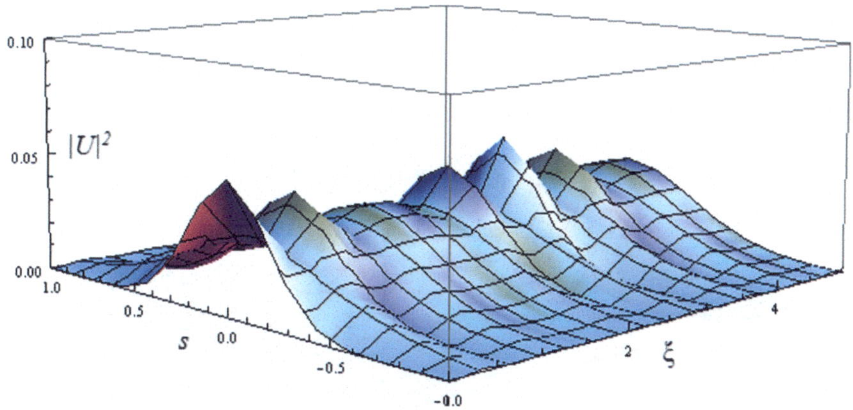

Fig. 7.9 Propagation of soliton when the waveguding parameter is augmented to $\delta = 10$. $P = 0.055$ and $r = 0.2831$. (Reprinted from Wave Motion, 77, Aavishkar Katti, R.A. Yadav, Awadhesh Prasad, Bright optical spatial solitons in photorefractive waveguides having both the linear and quadratic electro-optic effect, 64–76, Copyright 2018, with permission from Elsevier)

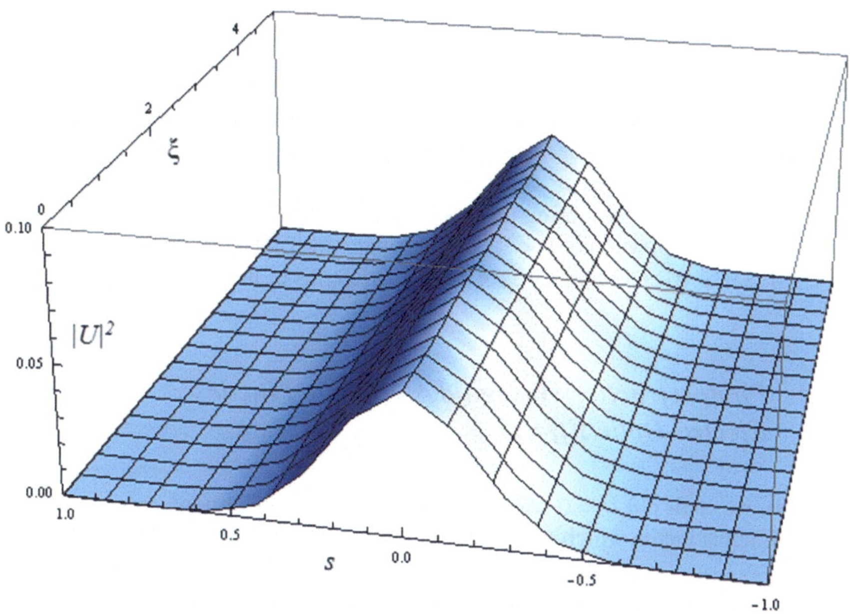

Fig. 7.10 Propagation of soliton when the waveguding parameter is augmented to $\delta = 30$. $P = 0.055$ and $r = 0.2831$. (Reprinted from Wave Motion, 77, Aavishkar Katti, R.A. Yadav, Awadhesh Prasad, Bright optical spatial solitons in photorefractive waveguides having both the linear and quadratic electro-optic effect, 64–76, Copyright 2018, with permission from Elsevier)

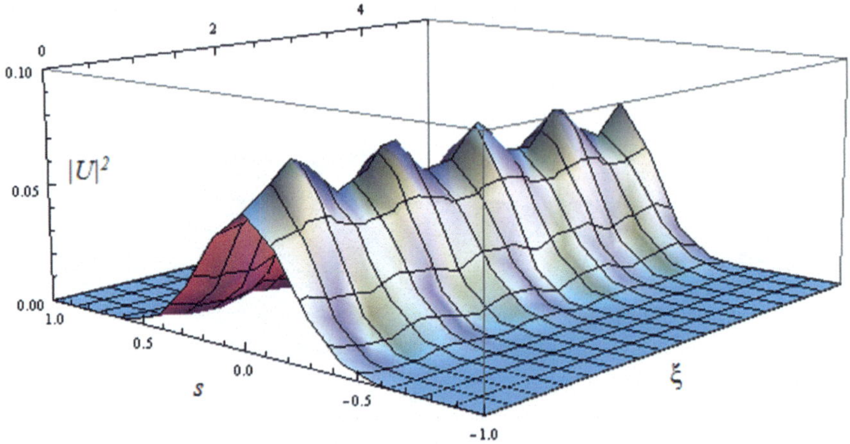

Fig. 7.11 Propagation of soliton when the waveguding parameter is augmented to $\delta = 50$. $P = 0.055$ and $r = 0.2831$. (Reprinted from Wave Motion, 77, Aavishkar Katti, R.A. Yadav, Awadhesh Prasad, Bright optical spatial solitons in photorefractive waveguides having both the linear and quadratic electro-optic effect, 64–76, Copyright 2018, with permission from Elsevier)

to increasing distance of propagation but the oscillation amplitude is less than one. This indicates very clearly that self trapping can occur even at peak powers much less than the threshold power due to the waveguiding effect of the embedded planar waveguide. With even greater values of δ, the power required to self trap the soliton reduces even more. Considering the case when the power of the light beam is less than the threshold power, i.e. $P < P_{0t1}$, the propagation of the light beam is shown in Figs. 7.8, 7.9, 7.10, 7.11 and 7.12 for various strengths of the waveguide.

7.3.2 Photorefractive Waveguides with the Linear Electro-Optic Effect

In (7.19), we shall put $b = 0$ since we consider conventional non centrosymmetric photorefractive media which does not have the quadratic electro-optic effect,

$$\frac{d^2 f(\xi)}{d\xi^2} = \frac{1}{r^4 f^3(\xi)} - 2\beta_1 \frac{\frac{P_0}{r^2 f^2(\xi)}}{\left(1 + \frac{P_0}{f(\xi)}\right)^2} - 2\delta f(\xi) \tag{7.21}$$

We have to look for points of equilibrium of (7.21) as a soliton forms when the beam width $f(\xi)$ remains constant. Hence, in (7.21), putting LHS equal to zero we obtain,

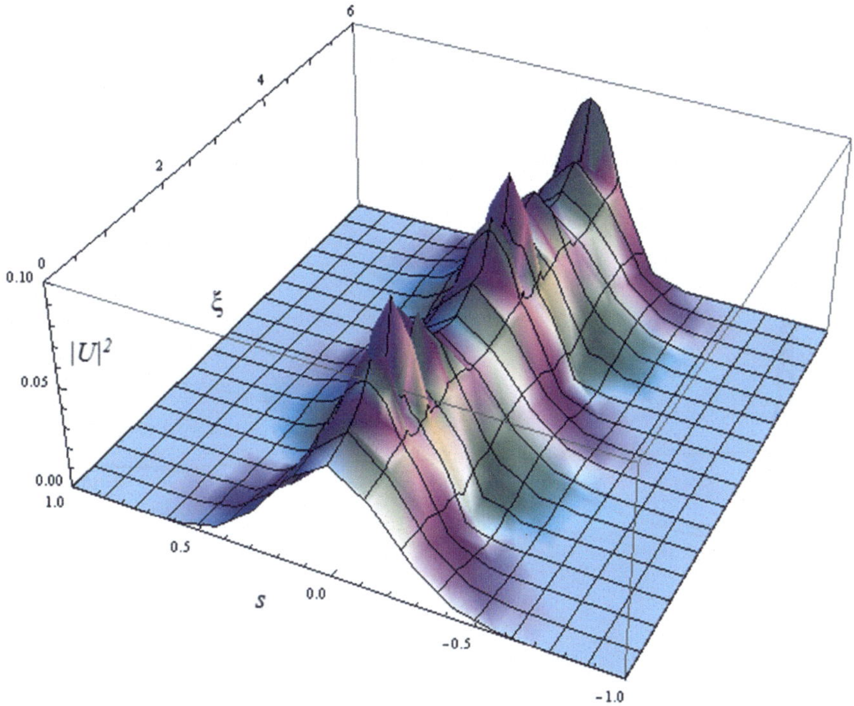

Fig. 7.12 Propagation of soliton when the waveguding parameter is augmented to $\delta = 100$. $P = 0.055$ and $r = 0.2831$. (Reprinted from Wave Motion, 77, Aavishkar Katti, R.A. Yadav, Awadhesh Prasad, Bright optical spatial solitons in photorefractive waveguides having both the linear and quadratic electro-optic effect, 64–76, Copyright 2018, with permission from Elsevier)

$$\frac{1}{r^4} = 2\delta + \frac{2\beta_1 P_{0t}}{r^2(1+P_{0t})^2} \tag{7.22}$$

(7.22) signifies an equilibrium condition containing four roots and hence which is quartic. The threshold power P_{0t} for stationary propagation of the light beam as a soliton can be inferred from (7.22) which serves to identify the existence for screening solitons in a photorefractive waveguide. As in the previous case, only one of the four solutions is real and positive so is physically acceptable.

We now need to consider a Lithium Niobate (LN) crystal for illustrating the characteristics of solitons in this case. The parameters taken in our investigation are shown clearly in Table 7.2. Bistable states can again be inferred from Fig. 7.13.

We shall proceed similarly to the previous case to predict the behavior of spatial solitons in absence of the waveguiding effect. For illustration, we take four different values of power, $P_1(= 0.0999) < P_{0t1}, P_2(= 0.3225) = P_{0t1}, P_{0t1} < P_3(= 1) < P_{0t2}$, $P_4(= 32.25) > P_{0t2}$. The variation in the variable beam width parameter with propagation is plotted in Fig. 7.14. As explained previously, we observe a self trapping if the power of the light beam falls within the two threshold powers. Self trapping

Table 7.2 Parameters for LN crystal taken in our theoretical investigation [13]

n_e	2.35	*V (bias emf)*	2000 V
x_0	40 μm	*l (crystal width)*	1 cm
λ_0	500 nm	ρ	0
r_{eff}	224×10^{-12} m/V	β_1	173

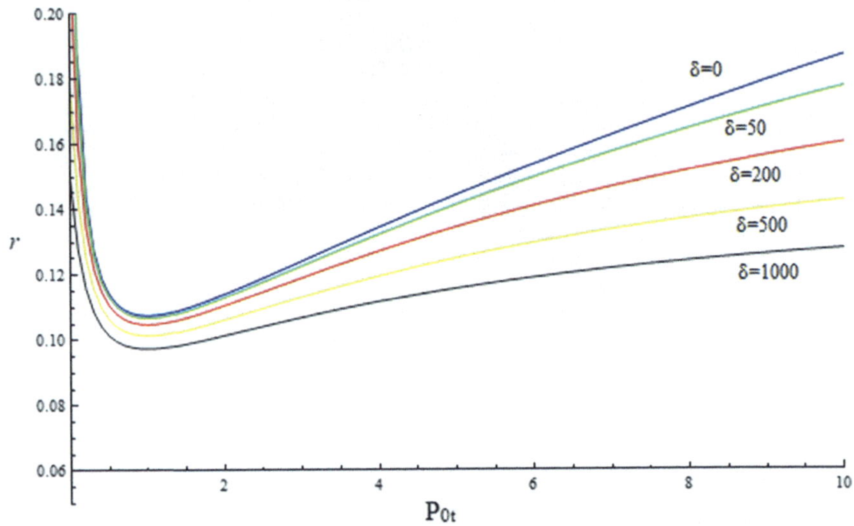

Fig. 7.13 Equilibrium spatial width r versus the threshold power P_{0t} of the spatial solitons, $\beta_1 = 173$. (Reprinted by permission from Springer Nature: Springer Optical and Quantum Electronics Bright screening solitons in a photorefractive waveguide, Aavishkar Katti, Copyright 2018)

is not observed for powers below the threshold power. The change induced by the embedded waveguide in the self trapping is investigated in Fig. 7.15 where the change in the variable beam width parameter is plotted with propagation considering diverse waveguide strengths. The conclusions remain same that a soliton of peak power lesser than the threshold power can still be self trapped by the use of an embedded waveguide in the photorefractive crystal.

7.3.3 Centrosymmetric Photorefractive Waveguides

In (7.19), we shall put $a = 0$ since we consider centrosymmetric non centrosymmetric photorefractive media which does not exhibit the linear electro-optic effect,

$$\frac{d^2 f(\xi)}{d\xi^2} = \frac{1}{r^4 f^3(\xi)} - 4\beta_2 \frac{\frac{P_0}{r^2 f^2(\xi)}}{\left(1 + \frac{P_0}{f(\xi)}\right)^3} - 2\delta f(\xi) \tag{7.23}$$

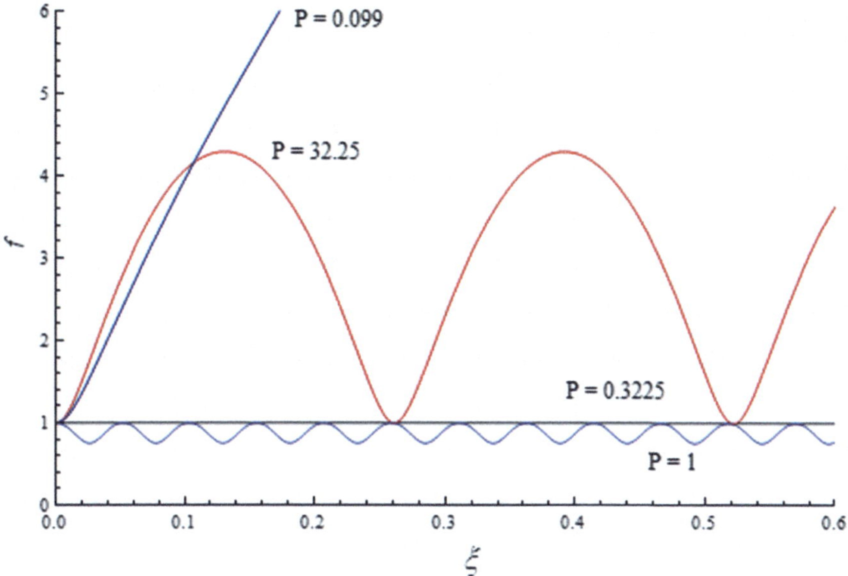

Fig. 7.14 Variable beam width parameter $f(\xi)$ versus the distance of propagation ξ for ddiverse values of soliton peak powers, $\beta_1 = 173$, $r = 0.125$. (Reprinted by permission from Springer Nature: Springer Optical and Quantum Electronics Bright screening solitons in a photorefractive waveguide, Aavishkar Katti, Copyright 2018)

We have to look for points of equilibrium of (7.23) as a soliton forms when the beam width $f(\xi)$ remains constant. Hence, in (7.23), putting LHS equal to zero, we obtain,

$$\frac{1}{r^4} = 2\delta + \frac{4\beta_2 P_{0t}}{r^2(1 + P_{0t})^3} \tag{7.24}$$

(7.24) signifies an equilibrium condition containing four roots and hence which is quartic. The threshold power P_{0t} for stationary propagation of the light beam as a soliton can be inferred from (7.24) which serves to identify the existence for screening solitons in a photorefractive waveguide. As in the previous case, only one of the four solutions is real and positive so is physically acceptable [7].

We now need to consider a Potassium Lithium Tantalate Niobate (KLTN) crystal for illustrating the characteristics of solitons in this case. Taking typical parameters [7], we get,$\beta_2 = 157.9$. The value of the equilibrium spatial width is plotted with respect to the threshold power in Fig. 7.16. Bistable states can again be inferred from Fig. 7.16.

We shall now proceed to predict the propagation behavior of spatial solitons when there is no waveguide. Consider four different values of power,$P_1(= 0.0999) < P_{0t1}$,$P_2(= 0.3341) = P_{0t1}$,$P_{0t1} < P_3(= 1) < P_{0t2}$, $P_4(= 33.41) > P_{0t2}$. We plot the

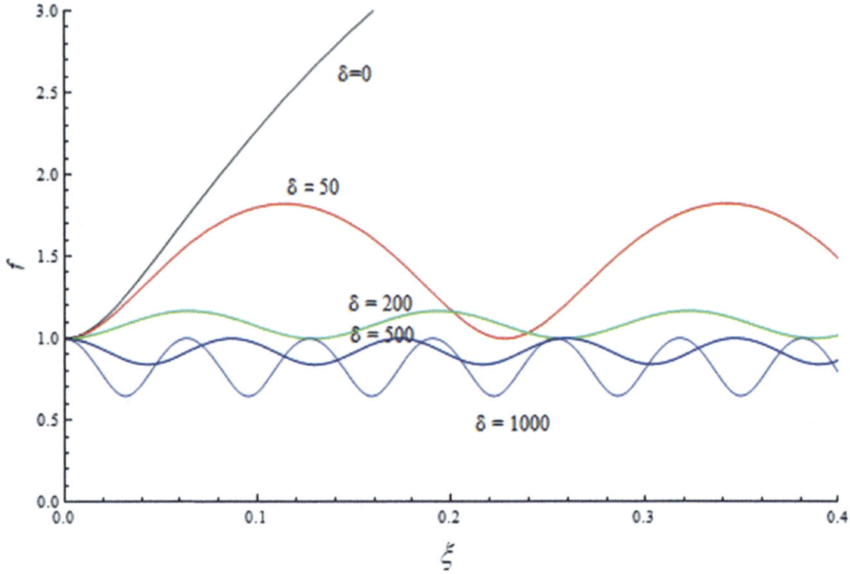

Fig. 7.15 Variable beam width parameter $f(\xi)$ versus distance of propagation ξ for different waveguide strengths, $\beta_1 = 173$, $P_0 = 0.075$ and $r = 0.161$. (Reprinted by permission from Springer Nature: Springer Optical and Quantum Electronics Bright screening solitons in a photorefractive waveguide, Aavishkar Katti, Copyright 2018)

Fig. 7.16 Equilibrium spatial width r as a function of the threshold peak power P_{0t} of the spatial solitons, $\beta_2 = 157.9$. (Reprinted by permission from Springer Nature: Springer Journal of Optics, Waveguiding effect on optical spatial solitons in centrosymmetric photorefractive materials, Binay P Akhouri et al, Copyright 2016)

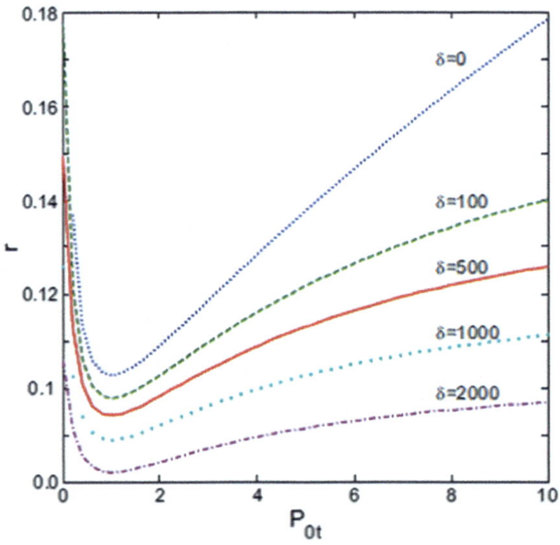

variation in f with the normalized propagation distance ξ in Fig. 7.17. As explained previously, we observe a self trapping if the power of the light beam falls within the two threshold powers. Self trapping is not observed for powers below the threshold power. For investigating the effect of waveguide embedded in the photorefractive crystal, we plot the beam width parameter as a function of the scaled distance of propagation for various strengths of the waveguide in Fig. 7.18. The conclusions remain same that a soliton of peak power lesser than the threshold power can still be self trapped by the use of an embedded waveguide in the photorefractive crystal.

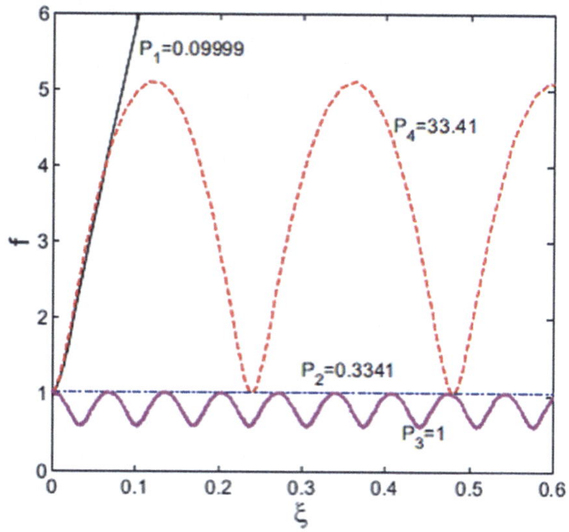

Fig. 7.17 Variable beam width parameter $f(\xi)$ versus distance of propagation ξ at four different soliton peak powers, $\beta = 157.9$. (Reprinted by permission from Springer Nature: Springer Journal of Optics, Waveguiding effect on optical spatial solitons in centrosymmetric photorefractive materials, Binay P Akhouri et al, Copyright 2016)

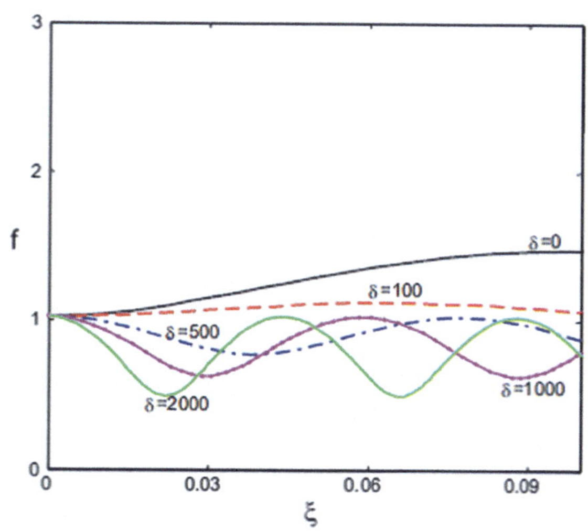

Fig. 7.18 Variable beam width parameter $f(\xi)$ versus the distance of propagation ξ at different waveguide strengths, $\beta = 157.9$, $P_0 = 0.06947$, $r = 0.1489$. (Reprinted by permission from Springer Nature: Springer Journal of Optics, Waveguiding effect on optical spatial solitons in centrosymmetric photorefractive materials, Binay P Akhouri et al, Copyright 2016)

7.3.4 Pyroelectric Photorefractive Waveguides

We shall now consider waveguide embedded in a photorefractive crystal with a finite pyroelectric coefficient. If the photorefractive crystal is unbiased, then we can say that the external bias has been replaced by the transient pyroelectric field which in turn induces a space charge field modulating the refractive index resulting in self trapping.

As before, we consider an optical beam propagating along the z-direction in a pyroelectric photorefractive waveguide embedded in a photorefractive crystal. The photorefractive crystal is placed between an insulating cover and a metallic plate whose temperature is precisely controlled by a Peltier cell. The optical c-axis of the photorefractive waveguide is along the x-direction and the beam is polarized along the x-direction. Also it is assumed diffraction is allowed along the x-direction only. Here the space charge field which is formed is due to the pyroelectric effect only and hence, $E_{sc} = E_{pysc}$. The slowly varying envelope of the electric field of the incident beam is expressed as, $\vec{E}(x, z) = \hat{x}\,\Phi(x, z)e^{ikz}$ where $\Phi(x, z)$ is the slowly varying envelope of the wave. Proceeding as previously, we obtain the dynamical evolution equation,

$$i\frac{\partial \Phi}{\partial z} + \frac{1}{2k_0 n_e}\frac{\partial^2 \Phi}{\partial x^2} - \frac{1}{2}k_0 n_e^3 r_{eff} E_{pysc}\Phi - gx^2\Phi = 0 \tag{7.25}$$

The induced space charge field in a photorefractive material due to exclusively the pyroelectric effect has been derived previously in Chap. 2,

$$E_{pysc} = -E_{py}\frac{I}{I + I_d} \tag{7.26}$$

where we also neglect the diffusion effects.

E_{py} is the transient pyroelectric field which is expressed as,

$$E_{py} = -\frac{1}{\epsilon_0 \epsilon_r}\frac{\partial P}{\partial T}\Delta T \tag{7.27}$$

In (7.27), $\frac{\partial P}{\partial T}$ is the pyroelectric coefficient, ϵ_0 is the vacuum permittivity, ΔT is the temperature change of the photorefractive crystal, ϵ_r is the dielectric constant. Using dimensionless coordinates mentioned before, the evolution equation becomes,

$$i\frac{\partial U}{\partial \xi} + \frac{1}{2}\frac{\partial^2 U}{\partial s^2} + \alpha\frac{|U|^2}{1 + |U|^2}U - \delta s^2 U = 0 \tag{7.28}$$

where,

$\alpha = \frac{(k_0 x_0)^2 n_e^4 r_{33}}{2} E_{py}$ and $\delta = g k_0 x_0^4 n_e$.

Equation (7.28) cannot be solved exactly, so we have to resort to the variational method using the paraxial approximation. The light beam envelope can be expressed as,

$$U(\xi, s) = U_0(\xi, s)e^{-i\,\Omega(\xi,s)} \tag{7.29}$$

$U_0(\xi, s)$ is a real quantity and $\Omega(\xi, s)$ represents the phase. Substituting (7.29) in (7.28) gives,

$$\left(i\frac{\partial U_0}{\partial \xi} + U_0\frac{\partial \Omega}{\partial \xi}\right) + \frac{1}{2}\left\{\frac{\partial^2 U_0}{\partial s^2} - 2i\frac{\partial U_0}{\partial s}\frac{\partial \Omega}{\partial s} - iU_0\frac{\partial^2 \Omega}{\partial s^2} - U_0\left(\frac{\partial \Omega}{\partial s}\right)^2\right\}$$
$$+\alpha\frac{U_0^2}{\left(1+U_0^2\right)}U_0 - \delta s^2 U_0 = 0 \tag{7.30}$$

The real and imaginary parts can be separately equated to give,

$$\frac{\partial U_0}{\partial \xi} - \frac{\partial U_0}{\partial s}\frac{\partial \Omega}{\partial s} - \frac{1}{2}U_0\frac{\partial^2 \Omega}{\partial s^2} = 0 \tag{7.31}$$

$$U_0\frac{\partial \Omega}{\partial \xi} + \frac{1}{2}\frac{\partial^2 U_0}{\partial s^2} - \frac{1}{2}U_0\left(\frac{\partial \Omega}{\partial s}\right)^2 + \alpha\Theta_1(\xi, s)U_0 - \delta s^2 U_0 = 0 \tag{7.32}$$

where

$$\Theta_1(\xi, s) = \frac{|U_0|^2}{1+|U_0|^2} \tag{7.33}$$

The last term in (7.32) refers to the contribution of the embedded planar waveguide. The last two terms represent the counteracting effect on diffraction leading to a stable self trapping.Again, following our previous analysis and taking the following Gaussian ansatz for a quasi soliton in our investigation,

$$U_0(\xi, s) = \frac{U_{00}}{\sqrt{f(\xi)}}e^{-s^2/2r^2 f^2(\xi)} \tag{7.34}$$

$$\Omega(\xi, s) = \frac{s^2}{2}\Gamma(\xi) \tag{7.35}$$

$$\Gamma(\xi) = -\frac{1}{f(\xi)}\frac{df(\xi)}{d\xi} \tag{7.36}$$

where the symbols have their usual meanings as defined in Sect. 7.3.1. We define the spatial width of the soliton as the product $rf(\xi)$. We term the solution (7.34) as a variational solution where the variable parameters to be found are are r, $f(\xi)$,

$\Gamma(\xi)$. In general, we shall assume that $\frac{df}{d\xi} = 0$ at $\xi = 0$, i.e., the soliton beam is non-diverging at the entry point of the crystal. Also, we can assume that $f = 1$ at $\xi = 0$. The next step is to simplify the non-linear term Θ_1. Expanding in a Taylor series to first order, we get,

$$\Theta_1(\xi, s) \approx \frac{\frac{U_{00}^2}{f(\xi)}}{\left(1 + \frac{U_{00}^2}{f}\right)} + s^2 \frac{\frac{U_{00}^2}{r^2 f^3(\xi)}}{\left(1 + \frac{U_{00}^2}{f(\xi)}\right)^2} \tag{7.37}$$

Substituting (7.34)–(7.37) in (7.28) results in an equation in several powers of s^2. Equating the coefficients of various powers of s^2, we obtain an evolution equation for $f(\xi)$,

$$\frac{d^2 f(\xi)}{d\xi^2} = \frac{1}{r^4 f^3(\xi)} - 2\alpha \frac{\frac{P_0}{r^2 f^2(\xi)}}{\left(1 + \frac{P_0}{f(\xi)}\right)^2} - 2\delta f(\xi) \tag{7.38}$$

The normalized power of the soliton is $P_0 = U_{00}^2$. Higher order terms of s, i.e., s^3, s^4 will not be considered since we take the first order approximation **[*r f(ξ)*] > *s*** in (7.37). Now, the light beam's propagation in the photorefractive crystal may proceed in the following three ways: it may travel stably with an unchanging intensity profile, it may diverge, or it may be compressed. From (7.38), we can see that this depends upon the magnitudes of the power P_0 and the parameters α. A soliton will be formed when the variable beam width parameter $f(\xi)$ remains unchanged with propagation. Hence equating the LHS in (7.38) to zero,

$$\frac{1}{r^4} = 2\delta + \frac{2\alpha P_{0t}}{r^2(1 + P_{0t})^2} \tag{7.39}$$

In (7.39), we see an equilibrium condition which is known as quartic equilibrium condition because it is a polynomial equation of fourth order and has four roots. From (7.39), we can find out the threshold power P_{0t} needed for stationary propagation of the optical beam. (7.39) is the existence equation for pyroelectric solitons travelling in a pyroelectric photorefractive waveguide. Examining the solutions of the above equation, two roots are imaginary, one root is negative and only one root is positive. Neglecting the imaginary and negative values of r for spatial solitons as being unphysical, we have one solution which remains physical.

For illustration of dynamical evolution of the solitons in such waveguides, SBN crystal will be considered which exhibits a strong pyroelectric effect. The parameters taken for the aforementioned crystal are shown in Table 7.3. Figure 7.19 shows the graph of r versus the threshold power P_{0t} for various strengths of the waveguide. We can see that the soliton width decreases with an increase in power for low powers, while the it increases with an increase in power for high powers. A similar dependence remains for each value of the waveguide parameter but the curve becomes less steep

Table 7.3 Parameters for SBN crystal taken in our theoretical investigation[4]

n_e	2.35	ΔT	20 °C
x_0	20 μm	ϵ_0	8.85×10^{-12} F/m
λ_0	532 nm	ϵ_r	3400
r_{eff}	237×10^{-12} m/V	α	40.2
$\frac{\partial P}{\partial T}$	-3×10^{-4} C m^{-2} K^{-1}		

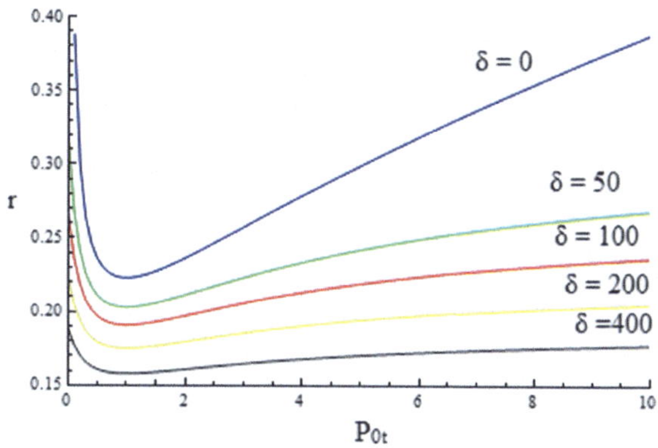

Fig. 7.19 The variation of equilibrium spatial width r with threshold peak power P_{0t} of the spatial solitons.$\alpha = 40.2$. (Reprinted from Optik - International Journal for Light and Electron Optics, 156, Aavishkar Katti, Bright pyroelectric quasi-solitons in a photorefractive waveguide, 433-438, Copyright 2018, with permission from Elsevier)

as the waveguide parameter increases. The existence of bistable states is clear from Fig. 7.19 since we can infer two values for the threshold powers for a single value of r and $f(\xi)$ is constant. The two values of the threshold power at which a soliton can just form are known as P_{0t1} and P_{0t2}. Also, the soliton width decreases with an increase in power in the low power region while the soliton width increases with an increase in power in the high power region.

It is worthwhile to first investigate the behaviour of the quasi-solitons when there is no waveguide present. As before, take four different values of power, below the first threshold power, at the first threshold power, between the two threshold powers and above the second threshold power, $P_1(= 0.0999) < P_{0t1}$, $P_2(= 0.2435) = P_{0t1}$, $P_{0t1} < P_3(= 1) < P_{0t2}$, $P_4(= 24.35) > P_{0t2}$. The change in f with the normalized propagation distance ξ is plotted in Fig. 7.20. To investigate the effect of the waveguide structure embedded in the photorefractive crystal, we plot the variation of the beam width parameter f with the normalized distance of propagation for different waveguide parameter δ in Fig. 7.21.

7.3.5 *Photovoltaic Photorefractive Waveguides*

If we consider the photorefractive crystal with a finite photovoltaic coefficient, the space charge field responsible for screening photovoltaic solitons has been found previously as,

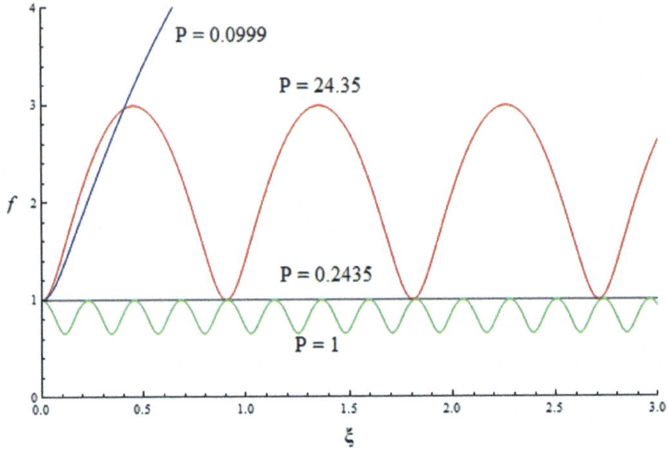

Fig. 7.20 The variation of variable beam width parameter $f(\xi)$ with normalized distance of propagation ξ at four different soliton peak powers $\alpha = 40.2$. (Reprinted from Optik - International Journal for Light and Electron Optics, 156, Aavishkar Katti, Bright pyroelectric quasi-solitons in a photorefractive waveguide, 433–438, Copyright 2018, with permission from Elsevier)

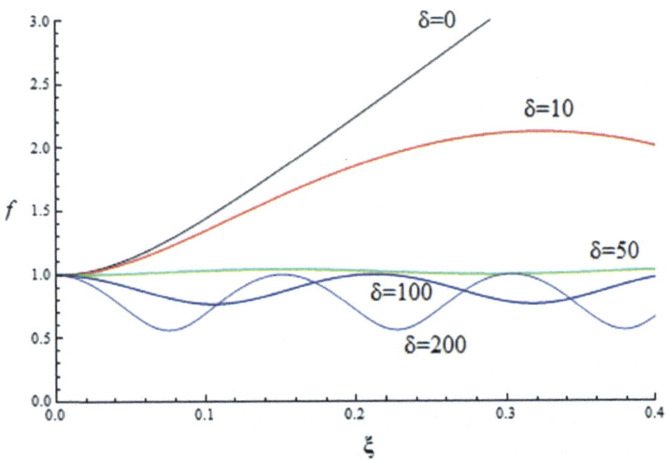

Fig. 7.21 Variable beam width parameter $f(\xi)$ versus the propagation distance ξ for different waveguide strengths, $\alpha = 40.2, P_0 = 0.055$, $r = 0.2810$. (Reprinted from Optik - International Journal for Light and Electron Optics, 156, Aavishkar Katti, Bright pyroelectric quasi-solitons in a photorefractive waveguide, 433–438, Copyright 2018, with permission from Elsevier)

$$E_{sc} = \frac{I_\infty + I_d}{I + I_d} E_0 + \frac{I_\infty - I_d}{I + I_d} E_p \tag{7.40}$$

Substituting this in (7.4) alongwith $b = 0$,

$$i\frac{\partial U}{\partial \xi} + \frac{1}{2}\frac{\partial^2 U}{\partial s^2} - \beta\left(\frac{1+\rho}{1+|U|^2}\right)U - \alpha\left(\frac{\rho-|U|^2}{1+|U|^2}\right)U - \delta s^2 U = 0 \qquad (7.41)$$

where,
$\beta_1 = a\frac{(k_0x_0)^2 n_e^4 r_{eff}}{2}E_0, \alpha = \frac{(k_0x_0)^2 n_e^4 r_{eff}}{2}E_p$ and $\delta = gk_0x_0^4 n_e$ alongwith $\rho = I_\infty/I_d$. Also $\rho = 0$ since bright solitons will be considered. The beam envelope in the slowly varying envelope approximation is expressed as [6],

$$U(\xi, s) = U_0(\xi, s)e^{-i\,\Omega(\xi,s)} \qquad (7.42)$$

$U_0(\xi, s)$ is the amplitude which is real and the phase is given by $\Omega(\xi, s)$. Substituting (7.42) in (7.41) we obtain,

$$\left(i\frac{\partial U_0}{\partial \xi} + U_0\frac{\partial \Omega}{\partial \xi}\right) + \frac{1}{2}\left\{\frac{\partial^2 U_0}{\partial s^2} - 2i\frac{\partial U_0}{\partial s}\frac{\partial \Omega}{\partial s} - iU_0\frac{\partial^2 \Omega}{\partial s^2} - U_0\left(\frac{\partial \Omega}{\partial s}\right)^2\right\}$$
$$- \beta_1\frac{1+\rho}{(1+U_0^2)}U_0 - \beta_2\left(\frac{1+\rho}{1+U_0^2}\right)^2 U_0 - \delta s^2 U_0 = 0 \qquad (7.43)$$

(7.43) is an equation which is a combination of real and imaginary terms. For the LHS to be zero, we need to equate the real and imaginary parts separately to zero,

$$\frac{\partial U_0}{\partial \xi} - \frac{\partial U_0}{\partial s}\frac{\partial \Omega}{\partial s} - \frac{1}{2}U_0\frac{\partial^2 \Omega}{\partial s^2} = 0 \qquad (7.44)$$

$$U_0\frac{\partial \Omega}{\partial \xi} + \frac{1}{2}\frac{\partial^2 U}{\partial s^2} - \frac{1}{2}U_0\left(\frac{\partial \Omega}{\partial s}\right)^2 - \beta_1\Phi_1(\xi, s)U_0 + \alpha\Phi_2(\xi, s)U_0 - \delta s^2 U_0 = 0$$
$$(7.45)$$

where,

$$\Phi_1(\xi, s) = \frac{1}{1+|U_0|^2} \qquad (7.46)$$

$$\Phi_2(\xi, s) = \frac{|U_0|^2}{(1+|U_0|^2)} \qquad (7.47)$$

$\Phi_1(\xi, s)$ and $\Phi_2(\xi, s)$ represent the contributions to the refractive index change due to the (linear) electro-optic effect. We shall be searching for physically acceptable bright soliton states and like before, assume a quasi-soliton solution for (7.45) as follows,

$$U_0(\xi, s) = \frac{U_{00}}{\sqrt{f(\xi)}}e^{-s^2/2r^2 f^2(\xi)} \qquad (7.48)$$

$$\Omega(\xi, s) = \frac{s^2}{2}\Gamma(\xi) \tag{7.49}$$

$$\Gamma(\xi) = -\frac{1}{f(\xi)}\frac{df(\xi)}{d\xi} \tag{7.50}$$

where $P_0 = U_{00}^2$ is the normalized peak power of the soliton, r is a constant which is always positive, $f(\xi)$ is the variable parameter related to the beam width such that the product $rf(\xi)$ is the spatial width of the soliton. The ansatz used in (7.48) can be said to be a variational solution. The variable parameters to be found are r, (ξ), $\Gamma(\xi)$. In general, we shall assume that the soliton beam is stable and not diverging when it enters the photorefractive crystal, i.e., $\frac{df}{d\xi} = 0$ at $\xi = 0$. Also, we can assume that $f = 1$ at $\xi = 0$. The next step is to simplify the non-linear terms Φ_1 and Φ_2. By a simple Taylor expansion in first order analogous to the calculation in Sect. 7.3.1, we get,

$$\Phi_1(\xi, s) \approx \frac{1}{\left(1 + \frac{U_{00}^2}{f}\right)} + s^2\frac{\frac{U_{00}^2}{r^2 f^3(\xi)}}{\left(1 + \frac{U_{00}^2}{f(\xi)}\right)^2} \tag{7.51}$$

$$\Phi_2(\xi, s) \approx \frac{\frac{U_{00}^2}{f}}{\left(1 + \frac{U_{00}^2}{f}\right)^2} - s^2\frac{\frac{U_{00}^2}{r^2 f^3(\xi)}}{\left(1 + \frac{U_{00}^2}{f(\xi)}\right)^2} \tag{7.52}$$

Substituting (7.14)–(7.18) in (7.11), we obtain an equation which contains several powers of s^2. Considering the coefficients of s^2 in LHS and RHS and equating them, we get the following evolution equation for the parameter $f(\xi)$,

$$\frac{d^2 f(\xi)}{d\xi^2} = \frac{1}{r^4 f^3(\xi)} - 2(\beta + \alpha)\frac{\frac{P_0}{r^2 f^2(\xi)}}{\left(1 + \frac{P_0}{f(\xi)}\right)^2} - 2\delta f(\xi) \tag{7.53}$$

Higher order terms of s, i.e., s^3, s^4 will not be considered since we take the first order approximation **$[rf(\xi)] > s$** in (7.51)–(7.52). Now, the optical beam may diverge, be compressed or travel stably after self trapping. From (7.53), we can see that this depends upon the magnitudes of the power P_0 and the parameters β_1, β_2. A soliton, which is a stable self trapped solitary wave will be formed when the beam width $f(\xi)$ remains exactly constant. So the LHS in (7.53) should equal zero,

$$\frac{1}{r^4} = 2\delta + \frac{2(\alpha + \beta)P_{0t}}{r^2(1 + P_{0t})^2} \tag{7.54}$$

In (7.54), we see an equilibrium condition which is known as quartic equilibrium condition because it is a polynomial equation of fourth order and has four roots. From

(7.54), we can find out the threshold power P_{0t} needed for stationary propagation of the optical beam. It can be said to be an existence equation for optical spatial solitons propagating through the photorefractive waveguide. Examining the solutions of the above equation, two roots are imaginary, one root is negative and only one root is positive. Neglecting the imaginary and negative values of r for spatial solitons as being unphysical, we are left with only one solution (Fig. 7.22).

We shall proceed similarly to the previous case to predict the behavior of spatial solitons in absence of the waveguiding effect. Considering typical parameters of the LN crystal with external bias field, we have, For illustration, we take four different values of power, $P_1(=0.0999) < P_{0t1}$, $P_2(=0.3341) = P_{0t1}$, $P_{0t1} < P_3(=1) < P_{0t2}$, $P_4(=33.41) > P_{0t2}$. We plot the variation in f with the normalized propagation distance ξ in Fig. 7.23. As explained previously, we observe a self trapping if the

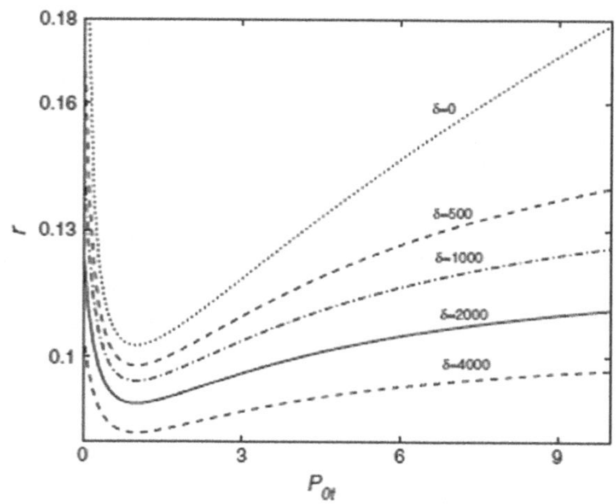

Fig. 7.22 Variation of equilibrium spatial width r with threshold peak power P_{0t} of the spatial solitons, $\beta = 157.9$, $\alpha = 31.58$. (Reprinted with permission from: S Shwetanshumala and S Konar, Bright optical spatial solitons in a photorefractive waveguide, Physica Scripta, 82, 4, 045404, First Published 2010-09-15, doi: 10.1088/0031-8949/82/04/045404)

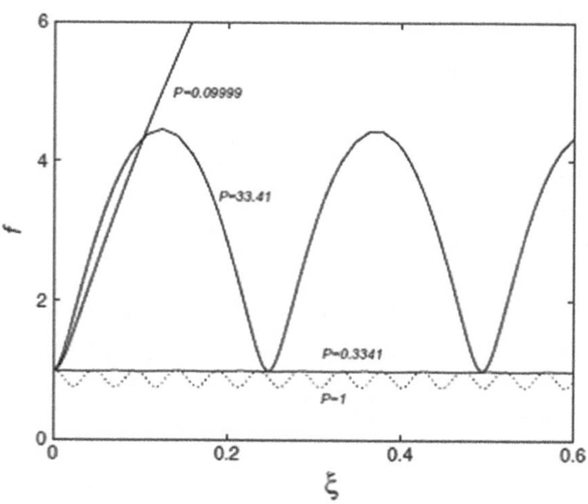

Fig. 7.23 Variable beam width parameter $f(\xi)$ versus the propagation distance ξ at four different soliton peak powers, $\beta = 157.9$, $\alpha = 31.58$. (Reprinted with permission from: S Shwetanshumala and S Konar, Bright optical spatial solitons in a photorefractive waveguide, Physica Scripta, 82, 4, 045404, First Published 2010-09-15, doi: 10.1088/0031-8949/82/04/045404)

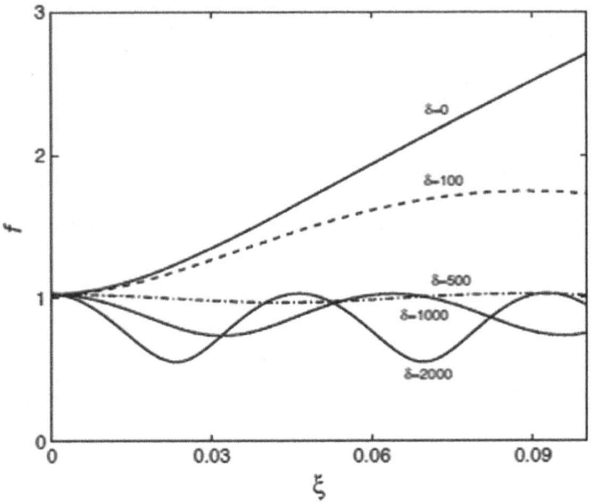

Fig. 7.24 The variation of variable beam width parameter $f(\xi)$ with of propagation distance ξ for different waveguide strengths, $\alpha = 31.58, \beta = 157.9$, $P_0 = 0.06947$, $r = 0.1489$. (Reprinted with permission from: S Shwetanshumala and S Konar, Bright optical spatial solitons in a photorefractive waveguide, Physica Scripta, 82, 4, 045404, First Published 2010-09-15, doi: 10.1088/0031-8949/82/04/045404)

power of the light beam falls within the two threshold powers. Self trapping is not observed for powers below the threshold power. For investigating the effect of waveguide embedded in the photorefractive crystal, we plot the change in the soliton width parameter f with the normalized distance of propagation for different waveguide parameters δ in Fig. 7.24. The conclusions remain same that a soliton of peak power lesser than the threshold power can still be self trapped by the use of an embedded waveguide in the photorefractive crystal.

7.4 Concluding Remarks

In conclusion, a comprehensive investigation of optical spatial solitons propagating in a photorefractive crystal having an embedded planar waveguide has been discussed. The paraxial diffraction equation is now modified to incorporate the waveguide term. In contrast to a numerical solution for the soliton enevelope, a modified Gaussian ansatz is used to solve this paraxial Helmholtz equation by means of a variational method. The planar waveguide structure increases the self-focusing while decreasing the minimum or threshold power required for self trapping. A soliton of lesser power which could not form can be formed with assistance from the self focusing due to the waveguide. The propagation of the solitons is visualized with and without the presence of the waveguide structure. The theory reduces to studying solitons in conventional photorefractive waveguides and centrosymmetric photorefractive waveguides. Lastly, this theory is extended for spatial solitons propagating in photovoltaic photorefractive crystals and pyroelectric photorefractive waveguides.

Note 1:

We shall take,

$$\vec{E} = \Phi(x, z)e^{ikz} \tag{7.55}$$

where the plane wave propagates along the z-axis and $\Phi(x, z)$ is the transverse modulation of the amplitude. Applying the paraxial approximation which implies a slow variation of $\Phi(x, z)$ with z compared to the wavelength.

The variation is expressed as, $\delta\Phi = \frac{\partial\Phi}{\partial z}\delta z \ll \Phi$ with $\delta z \sim \lambda$. So that

$$\frac{\partial\Phi}{\partial z} \ll \Phi/\lambda \sim k\,\Phi \tag{7.56}$$

Hence,

$$\frac{\partial^2\Phi}{\partial z^2} \ll k\frac{\partial\Phi}{\partial z} \ll k^2\Phi \tag{7.57}$$

Now, we have to evaluate,

$$\nabla^2\overrightarrow{E} + (k_0 n_e')^2\overrightarrow{E} - gx^2\overrightarrow{E} = 0 \tag{7.58}$$

The Laplacian can be expressed as a combination of the transverse and longitudinal parts,

$$\nabla^2 = \left[\nabla_\perp^2 + \partial_z^2\right] \tag{7.59}$$

So,

$$\nabla^2\vec{E} = \left[\nabla_\perp^2 + \partial_z^2\right]\vec{E} \tag{7.60}$$

Now,

$$\partial_z^2\vec{E} = \partial_z^2\left[\Phi(x, z)e^{ikz}\right] \tag{7.61}$$

Solving (7.61),

$$\partial_z^2\left[\Phi(x, z)e^{ikz}\right] = \left[\partial_z^2\Phi + 2ik\partial_z\Phi - k^2\Phi\right]e^{ikz} \tag{7.62}$$

Substitute in (7.58),

$$\nabla_{\perp}^{2}\left[\Phi(x,z)e^{ikz}\right] + \partial_{z}^{2}\Phi(x,z) + 2ik\partial_{z}\Phi(x,z) - k^{2}\Phi(x,z)]e^{ikz} + (k_{0}n_{e}')^{2}\Phi(x,z)e^{ikz} - gx^{2}\Phi(x,z)e^{ikz} \quad (7.63)$$

The term $\partial_{z}^{2}\Phi(x,z)$ can be neglected due to the paraxial approximation. So,

$$\nabla_{\perp}^{2}\left[\Phi(x,z)e^{ikz}\right] + 2ik\partial_{z}\Phi(x,z) - k^{2}\Phi(x,z)]e^{ikz} + (k_{0}n_{e}')^{2}\Phi(x,z)e^{ikz} - gx^{2}\Phi(x,z)e^{ikz} \quad (7.64)$$

From (7.2),

$$n_{e}'^{2} = n_{e}^{2} - n_{e}^{4}r_{33}E_{sc} - n_{e}^{4}g_{eff} \in_{0}^{2} (\in_{r} - 1)^{2}E_{sc}^{2} \quad (7.65)$$

Substitute $k = k_{0}n_{e}$ and (7.65) in (7.64),

$$i\frac{\partial\Phi}{\partial z} + \frac{1}{2k_{0}n_{e}}\left(\frac{\partial^{2}\Phi}{\partial x^{2}} + \frac{\partial^{2}\Phi}{\partial y^{2}}\right) - \frac{1}{2}k_{0}n_{e}^{3}r_{eff}E_{sc}\Phi - \frac{1}{2}k_{0}n_{e}^{3}g_{eff} \in_{0}^{2} (\in_{r} - 1)^{2}E_{sc}^{2}\Phi - gx^{2}\Phi = 0 \quad (7.66)$$

Since the diffraction effects are considered only in the in the x-direction, (7.66) becomes,

$$i\frac{\partial\Phi}{\partial z} + \frac{1}{2k_{0}n_{e}}\frac{\partial^{2}\Phi}{\partial x^{2}} - \frac{1}{2}k_{0}n_{e}^{3}r_{eff}E_{sc}\Phi - \frac{1}{2}k_{0}n_{e}^{3}g_{eff} \in_{0}^{2} (\in_{r} - 1)^{2}E_{sc}^{2}\Phi - gx^{2}\Phi = 0 \quad (7.67)$$

Note 2

We want to calculate,

$$\Phi_{1}(\xi,s) = \frac{1}{1 + |U_{0}|^{2}} \quad (7.68)$$

From (7.14),

$$U_{0}(\xi,s) = \frac{U_{00}}{\sqrt{f(\xi)}}e^{-s^{2}/2r^{2}f^{2}(\xi)} \quad (7.69)$$

Hence, we get,

$$|U_{0}|^{2} = \frac{U_{00}^{2}}{f(\xi)}e^{-s^{2}/r^{2}f^{2}(\xi)} \quad (7.70)$$

Now, expanding the exponential function in (7.70),

$$e^{-s^2/r^2 f^2(\xi)} = 1 - \frac{s^2}{r^2 f^2(\xi)} + \frac{1}{2!}\left(\frac{s^2}{r^2 f^2(\xi)}\right)^2 - \cdots \tag{7.71}$$

Substituting in (7.70),

$$|U_0|^2 = \frac{U_{00}^2}{f(\xi)}\left\{1 - \frac{s^2}{r^2 f^2(\xi)} + \frac{1}{2!}\left(\frac{s^2}{r^2 f^2(\xi)}\right)^2 - \cdots\right\} \tag{7.72}$$

Substitute (7.72) in (7.68), we have,

$$\Phi_1 = \frac{1}{1 + \frac{U_{00}^2}{f(\xi)}\left\{1 - \frac{s^2}{r^2 f^2(\xi)} + \frac{1}{2!}\left(\frac{s^2}{r^2 f^2(\xi)}\right)^2 - \cdots\right\}} \tag{7.73}$$

Simplifying,

$$\Phi_1 = \left[1 + \frac{U_{00}^2}{f(\xi)}\left\{1 - \frac{s^2}{r^2 f^2(\xi)} + \frac{1}{2!}\left(\frac{s^2}{r^2 f^2(\xi)}\right)^2 - \cdots\right\}\right]^{-1} \tag{7.74}$$

$$\Rightarrow \Phi_1 = \left[\left(1 + \frac{U_{00}^2}{f(\xi)}\right)\left\{1 - s^2\frac{\frac{U_{00}^2}{r^2 f^3(\xi)}}{\left(1 + \frac{U_{00}^2}{f(\xi)}\right)} + \frac{\frac{U_{00}^2}{f(\xi)}\frac{1}{2!}\left(\frac{s^2}{r^2 f^2(\xi)}\right)^2}{\left(1 + \frac{U_{00}^2}{f(\xi)}\right)} - \cdots\right\}\right]^{-1} \tag{7.75}$$

$$\Rightarrow \Phi_1 = \left(1 + \frac{U_{00}^2}{f(\xi)}\right)^{-1}\left\{1 - s^2\frac{\frac{U_{00}^2}{r^2 f^3(\xi)}}{\left(1 + \frac{U_{00}^2}{f(\xi)}\right)} + \frac{\frac{U_{00}^2}{f(\xi)}\frac{1}{2!}\left(\frac{s^2}{r^2 f^2(\xi)}\right)^2}{\left(1 + \frac{U_{00}^2}{f(\xi)}\right)} - \cdots\right\}^{-1} \tag{7.76}$$

Following the approach of Refs. [28, 29], the term in curly brackets is expanded in a Taylor series. Considering the first order approximation we get,

$$\Phi_1 \approx \left(1 + \frac{U_{00}^2}{f(\xi)}\right)^{-1}\left\{1 + s^2\frac{\frac{U_{00}^2}{r^2 f^3(\xi)}}{\left(1 + \frac{U_{00}^2}{f(\xi)}\right)}\right\} \tag{7.77}$$

$$\Phi_1(\xi, s) \approx \frac{1}{\left(1 + \frac{U_{00}^2}{f(\xi)}\right)} + s^2\frac{\frac{U_{00}^2}{r^2 f^3(\xi)}}{\left(1 + \frac{U_{00}^2}{f(\xi)}\right)^2} \tag{7.78}$$

Proceeding similarly for $\Phi_2(\xi, s) = \frac{1}{(1+|U_0|^2)^2}$, we get,

$$\Phi_2 = \frac{1}{\left(1 + \frac{U_{00}^2}{f(\xi)}\left\{1 - \frac{s^2}{r^2 f^2(\xi)} + \frac{1}{2!}\left(\frac{s^2}{r^2 f^2(\xi)}\right)^2 - \dots\right\}\right)^2} \tag{7.79}$$

Simplifying,

$$\Rightarrow \Phi_2 = \left(1 + \frac{U_{00}^2}{f(\xi)}\right)^{-2}\left\{1 - s^2\frac{\frac{U_{00}^2}{r^2 f^3(\xi)}}{\left(1 + \frac{U_{00}^2}{f(\xi)}\right)} + \frac{\frac{U_{00}^2}{f(\xi)}\frac{1}{2!}\left(\frac{s^2}{r^2 f^2(\xi)}\right)^2}{\left(1 + \frac{U_{00}^2}{f(\xi)}\right)} - \dots\right\}^{-2} \tag{7.80}$$

Expanding the second term in Taylor series in first order,

$$\Phi_2(\xi, s) \approx \frac{1}{\left(1 + \frac{U_{00}^2}{f(\xi)}\right)^2} + 2s^2\frac{\frac{U_{00}^2}{r^2 f^3(\xi)}}{\left(1 + \frac{U_{00}^2}{f(\xi)}\right)^3} \tag{7.81}$$

References

1. M. Segev, B. Crosignani, A. Yariv, B. Fischer, Spatial solitons in photorefractive media. Phys. Rev. Lett. **68**(7), 923–926 (1992). https://doi.org/10.1103/PhysRevLett.68.923
2. Z. Chen, M. Segev, D.N. Christodoulides, Optical spatial solitons: historical overview and recent advances. Rep. Prog. Phys. **75**(8), 086401 (2012). https://doi.org/10.1088/0034-4885/75/8/086401
3. A. Katti, Bright screening solitons in a photorefractive waveguide. Opt Quant Electron **50**(6), 263 (2018). https://doi.org/10.1007/s11082-018-1524-y
4. A. Katti, Bright pyroelectric quasi-solitons in a photorefractive waveguide. Optik—Int. J. Light Electron Opt. **156**, 433–438 (2018). https://doi.org/10.1016/j.ijleo.2017.10.105
5. A. Katti, R.A. Yadav, A. Prasad, Bright optical spatial solitons in photorefractive waveguides having both the linear and quadratic electro-optic effect. Wave Motion **77**, 64–76 (2018). https://doi.org/10.1016/J.WAVEMOTI.2017.10.002
6. S. Shwetanshumala, S. Konar, Bright optical spatial solitons in a photorefractive waveguide. Phys. Scr. **82**(4), 045404 (2010). https://doi.org/10.1088/0031-8949/82/04/045404
7. B.P. Akhouri, P.K. Gupta, Waveguiding effect on optical spatial solitons in centrosymmetric photorefractive materials. J. Opt. **46**(3), 281–286 (2017). https://doi.org/10.1007/s12596-016-0372-z
8. A. Katti, R.A. Yadav, D.P. Singh, Theoretical investigation of incoherently coupled solitons in centrosymmetric photorefractive crystals. Optik—Int. J. Light Electron Opt. **136**, 89–106 (2017). https://doi.org/10.1016/j.ijleo.2017.01.099
9. S.N. Vlasov, V.A. Petrishchev, V.I. Talanov, Averaged description of wave beams in linear and nonlinear media (the method of moments). Radiophys. Quantum Electron. **14**(9), 1062–1070 (1974). https://doi.org/10.1007/BF01029467
10. S.A. Akhmanov, A.P. Sukhorukov, R.V. Khokhlov, Self-focusing and difraction of light beams in a nonlinear medium. Soviet Physics Uspekhi **10**(5), 609–636 (1968). https://doi.org/10.1070/PU1968v010n05ABEH005849

11. M. Segev, G.C. Valley, B. Crosignani, P. Diporto, A. Yariv, Steady-state spatial screening solitons in photorefractive materials with external applied field. Phys. Rev. Lett. **73**(24), 3211 (1994)
12. D. Anderson, Variational approach to nonlinear pulse propagation in optical fibers. Phys. Rev. A **27**(6), 3135–3145 (1983). https://doi.org/10.1103/PhysRevA.27.3135
13. D.N. Christodoulides, M.I. Carvalho, Bright, dark, and gray spatial soliton states in photorefractive media. J. Opt. Soc. Am. B **12**(9), 1628 (1995). https://doi.org/10.1364/JOSAB.12.001628

The manufacturer's authorised representative in the EU is Springer Nature Customer Service Centre GmbH, Europaplatz 3, 69115 Heidelberg, Germany. If you have any concerns regarding our products, please contact ProductSafety@springernature.com

Printed and bound by CPI Group (UK) Ltd, Croydon, CR0 4YY
01/10/2026
02226294-0003